Mechanical sci
for technicians

Ian McDonagh
Senior lecturer in engineering,
Carlett Park College of Technology

Edward Arnold

First published 1979
by Edward Arnold (Publishers) Ltd
41 Bedford Square, London WC1B 3DQ

ISBN 0 7131 3411 9

British Library Cataloguing in Publication Data

McDonagh, I.
Mechanical science for technicians.
1. Mechanics, Applied
I. Title
620.1 TA350

ISBN 0-7131-3411-9

Text set in 10/11pt IBM Press Roman, printed by photolithography, and bound in Great Britain at the Pitman Press, Bath.

Contents

Preface

This book is designed to meet the requirements of the Technician Education Council (TEC) standard unit Mechanical Science III (U75/058) for mechanical and production engineering technicians, the aim of which is to develop the student's analytical techniques in the application of scientific principles to mechanical engineering.

SI units have been used throughout in the text, with the following preferred multiples and submultiples:

Prefix	*Symbol*	*Multiplication factor*
giga	G	$10^9 = 1\,000\,000\,000$
mega	M	$10^6 = 1\,000\,000$
kilo	k	$10^3 = 1000$
milli	m	$10^{-3} = 0.001$
micro	μ	$10^{-6} = 0.000\,001$

Ian McDonagh

1 Stress, strain, and elasticity

1.1 Types of force

There are three types of force which may be applied to a material:

i) tensile (or stretching) force, fig. 1.1(a);
ii) compressive (or squeezing) force, fig. 1.1(b);
iii) shear (or sliding) force, fig. 1.1(c).

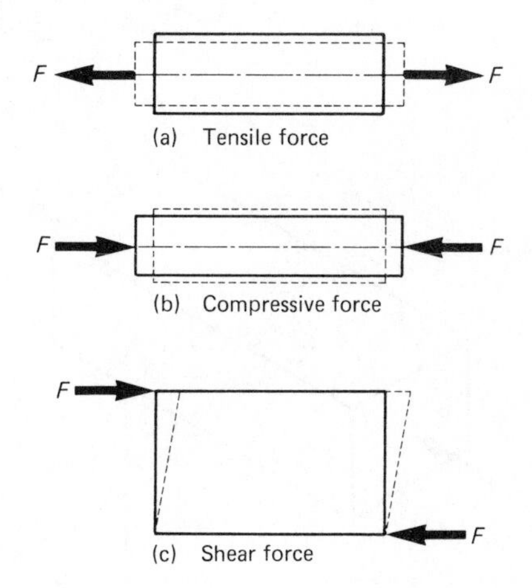

Fig. 1.1 Types of force

Tensile and compressive forces are *direct* forces. Direct forces are also known as *uniaxial* forces, since the opposing forces are in line, and these induce *direct* stresses in the material (see section 1.2). Shear forces are *indirect*, since the lines of action of the opposing forces must be separated as shown in fig. 1.1(c) for shear to occur. Shear forces induce *shear* stresses in the material (see section 1.8).

1.2 Direct stress

Direct stress is defined as the applied force F per unit cross-sectional area A resisting the force,

i.e. $$\text{direct stress} = \frac{\text{applied force}}{\text{cross-sectional area resisting the force}}$$

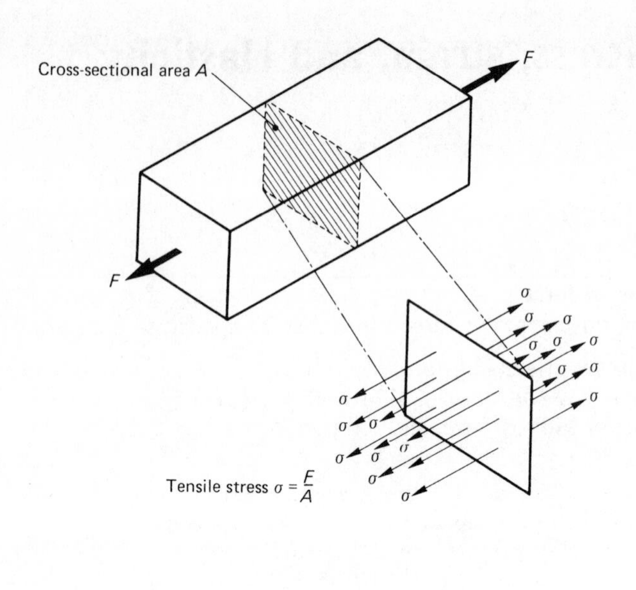

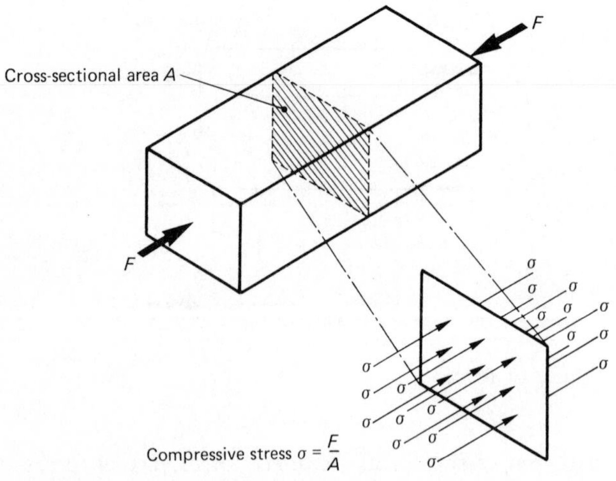

Fig. 1.2 Tensile and compressive stress

Direct stresses may be tensile or compressive, as illustrated in fig. 1.2. The symbol used for direct stress is σ (*sigma*); thus

$$\text{direct stress } \sigma = \frac{F}{A}$$

which should be remembered.

The basic unit for stress is the newton per square metre (N/m^2). Since this is very small, the unit meganewton per square metre (MN/m^2) is often used, and the units newton per square millimetre (N/mm^2) and *pascal* (Pa) may also be used. It is useful to remember that

$$1\ N/mm^2 = 1\ MN/m^2$$

and

$$1\ Pa = 1\ N/m^2$$

1.3 Strain

Strain is defined as change in dimension (x) per unit original dimension (l),

i.e. $$\text{strain} = \frac{\text{change in dimension}}{\text{original dimension}}$$

Strain may be tensile, compressive, or shear. Tensile strain occurs when there is an *increase* in the original dimension, and compressive strain when there is a *decrease*. Shear strain is discussed in section 1.9.

The symbol used for tensile or compressive strain is ϵ (*epsilon*),

$$\therefore \quad \epsilon = \frac{x}{l}$$

which should be remembered.

Since strain is a ratio of *like* quantities, *it has no units.*

1.4 Modulus of elasticity

All solid materials will change shape slightly when stressed. If a material reverts back to its original shape and size when the stress is removed, it is an *elastic* material. Most solid materials are elastic up to a certain stress limit known as the *elastic limit* – a common exception to this is lead at room temperature. The stress-strain graph for an elastic material in fig. 1.3 shows

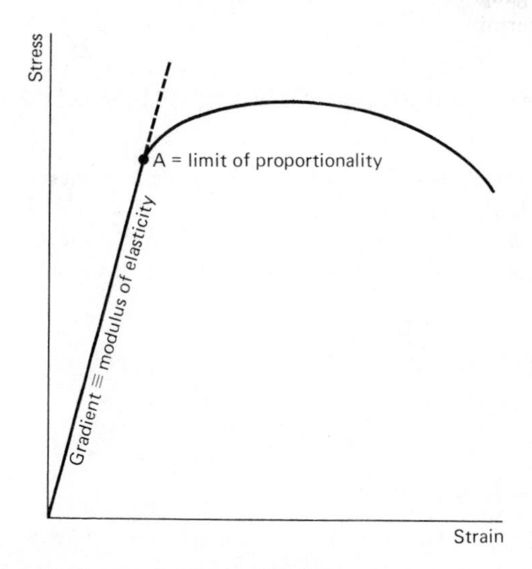

Fig. 1.3 Stress–strain graph for a material

that, up to a point which may be at or just below the elastic limit, the graph is a straight line; i.e., up to the point A (the *limit of proportionality*) in fig. 1.3, the stress is directly proportional to the strain, or

$$\sigma \propto \epsilon$$

$$\therefore \quad \frac{\sigma}{\epsilon} = \text{constant}$$

This constant is known as the *modulus of elasticity* or *Young's modulus*, *E*, for the material,

$$\text{i.e.} \quad \text{modulus of elasticity} = \frac{\text{stress}}{\text{strain}}$$

$$\text{or} \quad E = \frac{\sigma}{\epsilon}$$

which should be remembered.

The basic unit for E is the same as for stress, i.e. the newton per square metre (N/m^2), though the multiple giganewton per square metre (GN/m^2) is often used.

Table 1.1 shows typical values of the modulus of elasticity for various materials.

Material	*Modulus of elasticity* (N/m^2)
Carbon steel	210×10^9
Copper	120×10^9
Cast iron	100×10^9
Brass	90×10^9
Aluminium alloy	90×10^9

Table 1.1 Typical values of the modulus of elasticity. (Note that the modulus of elasticity of carbon steel is little altered for variations in the carbon content.)

Example In a tensile test on a steel sample, an extensometer recorded an increase in length of 0.117 mm for an applied load of 60 kN. If the diameter and the original gauge length of the sample were 15.96 mm and 80 mm respectively, determine the modulus of elasticity for the steel.

$$E = \sigma/\epsilon$$

$$\text{but} \quad \sigma = F/A \quad \text{and} \quad \epsilon = x/l$$

$$\therefore \quad E = \frac{Fl}{Ax}$$

which it is useful to remember.

Here, $F = 60\text{ kN} = 60 \times 10^3\text{ N}$ $\quad l = 80\text{ mm}$ $\quad x = 0.117\text{ mm}$

and $A = (\pi/4) \times (15.96 \times 10^{-3}\text{m})^2 = 200 \times 10^{-6}\text{m}^2$

$$\therefore \quad E = \frac{60 \times 10^3 \text{ N} \times 80 \text{ mm}}{200 \times 10^{-6} \text{ m}^2 \times 0.117 \text{ mm}}$$

$$= 205.1 \times 10^9 \text{ N/m}^2$$

i.e. the modulus of elasticity of the steel is 205.1 GN/m^2.

1.5 Stress and strain in composite bars

Consider a composite bar consisting of a steel tube which is completely filled with rubber, as shown in fig. 1.4(a). If the composite bar is subjected to a compressive force F applied through the flat plates on the ends of the tube, as shown in fig. 1.4(b), then the *whole* assembly will suffer a *decrease* in length x, the decrease being limited by the stiffness of the stronger material, i.e. the steel tube. Also, the compressive force F is *shared* between the tube and the rubber, with the tube, being the more rigid, carrying most of the force. Thus, to solve problems relating to composite bars of this type, it should be remembered that

a) decrease in length of tube = decrease in length of rubber

or compressive strain in tube = compressive strain in rubber

b) total force = force on tube + force on rubber

Let subscript a refer to the tube and subscript b to the rubber, then

$$\epsilon_a = \epsilon_b \quad \text{and} \quad F = F_a + F_b$$

But strain $\epsilon = \sigma/E$ and force $F = \sigma A$

$$\therefore \quad \frac{\sigma_a}{E_a} = \frac{\sigma_b}{E_b} \qquad \text{(i)}$$

$$\text{and} \quad F = \sigma_a A_a + \sigma_b A_b \qquad \text{(ii)}$$

which are useful to remember.

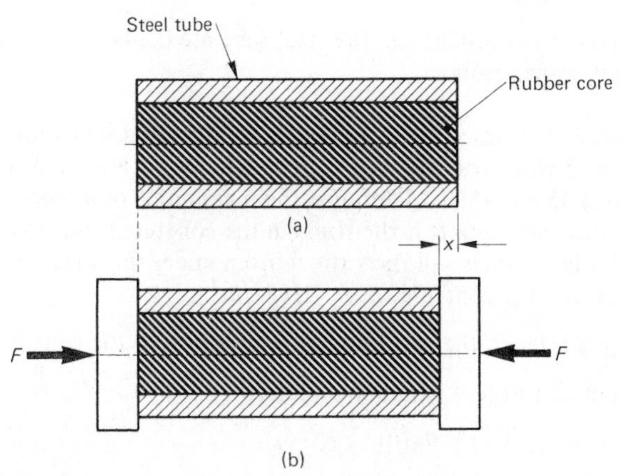

Fig. 1.4 Composite bar

Example 1 If the internal and external diameters of the steel tube shown in fig. 1.4 are 24 mm and 30 mm respectively, determine the compressive stresses in the rubber and in the steel when the applied force is 3 kN. For steel, $E = 200\ \text{GN/m}^2$; for rubber, $E = 2.5\ \text{GN/m}^2$.

Let subscript a refer to the steel tube and subscript b to the rubber.

$$\frac{\sigma_a}{E_a} = \frac{\sigma_b}{E_b}$$

$$\therefore \quad \sigma_a = \sigma_b\,(E_a/E_b)$$

where $E_a = 200\ \text{GN/m}^2$ and $E_b = 2.5\ \text{GN/m}^2$

$$\therefore \quad \sigma_a = \sigma_b \times \frac{200\ \text{GN/m}^2}{2.5\ \text{GN/m}^2} = 80\,\sigma_b$$

Also, $F = \sigma_a A_a + \sigma_b A_b = 80\,\sigma_b A_a + \sigma_b A_b$

where $F = 3\text{kN} = 3 \times 10^3\ \text{N}$

$$A_a = (\pi/4) \times (30^2 - 24^2)\text{mm}^2 = 254.5\ \text{mm}^2$$

and $A_b = (\pi/4) \times (24\ \text{mm})^2 = 452.4\ \text{mm}^2$

$$\therefore \quad 3 \times 10^3\ \text{N} = 80\,\sigma_b \times 254.5\ \text{mm}^2 + \sigma_b \times 452.4\ \text{mm}^2$$

$$= 2.081 \times 10^4\ \sigma_b\ \text{mm}^2$$

$$\therefore \quad \sigma_b = \frac{3 \times 10^3\ \text{N}}{2.081 \times 10^4\ \text{mm}^2}$$

$$= 0.144\ \text{N/mm}^2 \quad \text{or} \quad 0.144\ \text{MN/m}^2$$

$$\sigma_a = 80\,\sigma_b$$

$$\therefore \quad \sigma_a = 80 \times 0.144\ \text{MN/m}^2$$

$$= 11.52\ \text{MN/m}^2$$

i.e. the stresses in the rubber and the steel tube are $0.144\ \text{MN/m}^2$ and $11.52\ \text{MN/m}^2$ respectively.

Example 2 A cast-iron tube, 3 m long, is completely filled with concrete and used as a vertical strut. If the external diameter of the tube is 450 mm and the wall is 35 mm thick, determine the maximum compressive load the composite strut can support if the stress in the concrete is not to exceed $2\ \text{N/mm}^2$. By how much will the strut shorten under this load? For cast iron, $E = 100\ \text{GN/m}^2$; for concrete, $E = 10\ \text{GN/m}^2$.

Let subscript a refer to the cast iron and subscript b to the concrete; then,

strain in cast iron = strain in concrete

or $$\sigma_a/E_a = \sigma_b/E_b$$

$$\therefore \quad \sigma_a = \sigma_b\,(E_a/E_b)$$

where $\sigma_b = 2\,N/mm^2 \quad E_a = 100\,GN/m^2 \quad$ and $\quad E_b = 10\,GN/m^2$

$$\therefore \quad \sigma_a = 2\,N/mm^2 \times \frac{100\,GN/m^2}{10\,GN/m^2}$$

$$= 20\,N/mm^2$$

Also, total force = force on cast iron + force on concrete

or $$F = \sigma_a A_a + \sigma_b A_b$$

where $A_a = (\pi/4) \times (450^2 - 380^2)\,mm^2 = 45.6 \times 10^3\,mm^2$

and $A_b = (\pi/4) \times (380\,mm)^2 = 113.4 \times 10^3\,mm^2$

$$\therefore \quad F = (20\,N/mm^2 \times 45.6 \times 10^3\,mm^2) + (2\,N/mm^2 \times 113.4 \times 10^3\,mm^2)$$

$$= 1.14 \times 10^6\,N \quad \text{or} \quad 1.14\,MN$$

i.e. the maximum force the composite strut can support is 1.14 MN.

Strain $\epsilon = x/l = \sigma/E$

$$\therefore \quad x = l\,(\sigma/E)$$

where $l = 3\,m$ and, for concrete, $\sigma = 2\,N/mm^2 = 2 \times 10^6\,N/m^2$

and $E = 10 \times 10^9\,N/m^2$

$$\therefore \quad x = 3\,m \times \frac{2 \times 10^6\,N/m^2}{10 \times 10^9\,N/m^2}$$

$$= 6 \times 10^{-4}\,m \quad \text{or} \quad 0.6\,mm$$

i.e. the composite strut will shorten by 0.6 mm.

Example 3 A concrete column of square cross-section, 250 mm x 250 mm, is required to support an axial load of 875 kN. Determine the minimum number of steel rods, each of diameter 6 mm, which would be required to reduce the stress in the concrete to 8 N/mm². For steel, $E = 200\,GN/m^2$; for concrete, $E = 12\,GN/m^2$.

Let subscript a refer to the steel, subscript b refer to the concrete, and n be the number of steel rods required.

Strain in steel = strain in concrete

or $$\sigma_a/E_a = \sigma_b/E_b$$

$$\therefore \quad \sigma_a = \sigma_b\,(E_a/E_b)$$

where $\sigma_b = 8\,N/mm^2 \quad E_a = 200\,GN/m^2 \quad$ and $\quad E_b = 12\,GN/m^2$

$$\therefore \quad \sigma_a = 8\,N/mm^2 \times \frac{200\,GN/m^2}{12\,GN/m^2}$$

$$= 133.3\,N/mm^2$$

Total force = force on steel + force on concrete

or $$F = \sigma_a A_a + \sigma_b A_b$$

where $F = 875\text{ kN} = 875 \times 10^3\text{ N}$

$A_a = (\pi/4) \times (6\text{ mm})^2 n = (28.27\,n)\text{ mm}^2$

and $A_b = (250\text{ mm})^2 - A_a = (62.5 \times 10^3 - 28.27\,n)\text{ mm}^2$

$\therefore\ 875 \times 10^3\text{ N} = 133.3\text{ N/mm}^2 \times (28.27\,n)\text{ mm}^2$

$+ 8\text{ N/mm}^2 \times (62.5 \times 10^3 - 28.27\,n)\text{ mm}^2$

$\therefore\ 875 \times 10^3 = 3768.4\,n + 500 \times 10^3 - 226.2\,n$

from which, $n = 105.9$

i.e. the minimum number of steel rods required is 106.

1.6 Stresses due to temperature change

A solid material expands when its temperature is increased and contracts when its temperature is decreased. The change in dimension, x, which occurs with change in temperature is given by

$$x = l\,\alpha\,\Delta\theta$$

where l = original dimension,
α = coefficient of linear expansion,
and $\Delta\theta$ = *change* in temperature.

If a solid material is subjected to an increase in temperature $\Delta\theta$, and the resulting expansion is completely (or partially) restricted, a *compressive* stress will be induced in the material. Similarly, if there is a decrease in temperature and the resulting contraction is restricted, a *tensile* stress will be induced.

Figure 1.5(a) shows a rod with initial length l. If the temperature of the rod is increased from θ_1 to θ_2, the length of the rod will then be $(l + x)$, fig. 1.5(b), and the rod will remain unstressed. If the expansion is restricted as shown in fig. 1.5(c), then a compressive stress σ together with its associated strain will be induced in the rod, i.e. the *change* in temperature will *change* the stress from zero in fig. 1.5(a) to σ in fig. 1.5(c), provided that the expansion is restricted.

Referring to fig. 1.5(c), where the expansion is completely restricted,

$$\text{compressive strain } \epsilon = \frac{x}{l+x} = \frac{\sigma}{E}$$

$$\therefore \quad \sigma = \frac{Ex}{l+x}$$

But $x = l\alpha(\theta_2 - \theta_1) = l\alpha\Delta\theta$

$$\therefore \quad \sigma = \frac{El\alpha\Delta\theta}{l + l\alpha\Delta\theta} = \frac{E\alpha\Delta\theta}{1 + \alpha\Delta\theta}$$

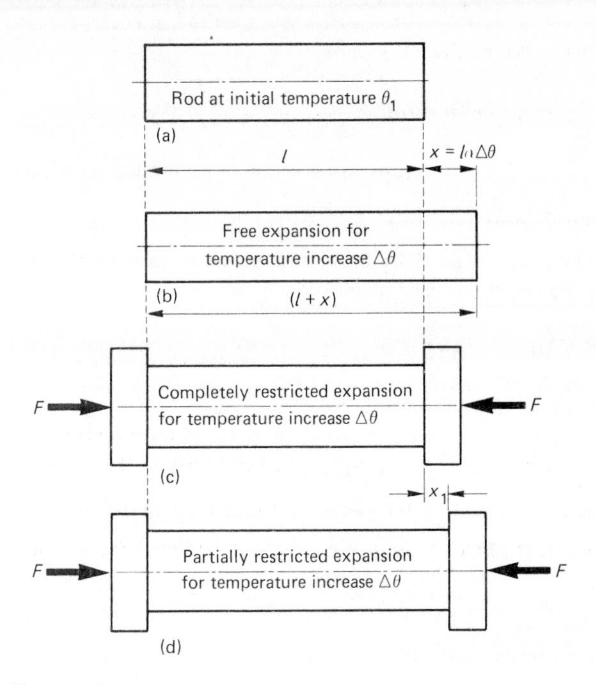

Fig. 1.5 Temperature stress

Since α is very small (typically 0.000 01), the term $\alpha\Delta\theta$ will be negligible compared with 1. Thus the term $(1 + \alpha\Delta\theta) \approx 1$; i.e., if the expansion is completely restricted, then a change in temperature $\Delta\theta$ will produce a change in stress of magnitude

$$\sigma = E\alpha\Delta\theta$$

which should be remembered.

Changes in stress which are induced by changes in temperature are known as temperature stresses.

Notice that $\alpha\Delta\theta = \sigma/E$

i.e. $\alpha\Delta\theta =$ *temperature strain*

which it is useful to remember.

If the expansion is restricted to an amount x_1 as shown in fig. 1.5(d), then

$$\text{total direct strain} = \frac{x - x_1}{l + x}$$

$$= \frac{l\alpha\Delta\theta - x_1}{l + l\alpha\Delta\theta}$$

$$= \frac{l\,[\alpha\Delta\theta - (x_1/l)]}{l\,(1 + \alpha\Delta\theta)}$$

As before, $(1 + \alpha\Delta\theta) \approx 1$

$$\therefore \quad \text{total direct strain} = \alpha\Delta\theta - \frac{x_1}{l}$$

$$= \text{temperature strain} - \text{strain due to stress}$$

But strain $= \sigma/E$

so, if the expansion is partially restricted, a change in temperature $\Delta\theta$ will produce a change in stress of magnitude

$$\sigma = E\left(\alpha\Delta\theta - \frac{x_1}{l}\right)$$

which should be remembered.

If a material is subjected to both an increase in temperature and a tensile stress (or a decrease in temperature and a compressive stress), then

total direct strain = temperature strain + strain due to stress

or total direct strain = sum of strains due to temperature change and stress

i.e.
$$\epsilon = \alpha\Delta\theta \pm \frac{\sigma}{E}$$

which should be remembered.

Example 1 The temperature of a steel bush which is initially unstressed is increased from 20°C to 80°C. If the expansion is completely restricted, determine the compressive stress in the bush material. Take $E = 200\,\text{GN/m}^2$ and $\alpha = 12 \times 10^{-6}/°\text{C}$.

Since the expansion is completely restricted, the increase in temperature will induce a change in stress of magnitude

$$\sigma = E\alpha\Delta\theta$$

where $E = 200\,\text{GN/m}^2 = 200 \times 10^9\,\text{N/m}^2$ $\qquad \alpha = 12 \times 10^{-6}/°\text{C}$

and $\Delta\theta = 80°\text{C} - 20°\text{C} = 60°\text{C}$

$$\therefore \quad \sigma = 200 \times 10^9\,\text{N/m}^2 \times 12 \times 10^{-6}/°\text{C} \times 60°\text{C}$$

$$= 144 \times 10^6\,\text{N/m}^2 \quad \text{or} \quad 144\,\text{MN/m}^2$$

i.e. the compressive stress in the bush material at 80°C is $144\,\text{MN/m}^2$.

Example 2 A spacer in a machine assembly was measured and found to be 200.00 mm long at a temperature of 20°C. After a period of time, the temperature of the spacer was found to be 60°C and its length 200.04 mm. Determine the compressive stress in the spacer at 60°C, if $E = 200\,\text{GN/m}^2$, $\alpha = 11.5 \times 10^{-6}/°\text{C}$, and the spacer was initially unstressed.

For partially restricted expansion, the increase in temperature will induce a change in stress of magnitude

$$\sigma = E(\alpha\Delta\theta - x_1/l)$$

where $E = 200\ \text{GN/m}^2 = 200 \times 10^9\ \text{N/m}^2$ $\quad \alpha = 11.5 \times 10^{-6}/°\text{C}$

$x_1 = 0.04\ \text{mm}$ $\quad l = 200\ \text{mm}$ and $\Delta\theta = 60°\text{C} - 20°\text{C} = 40°\text{C}$

$$\therefore \quad \sigma = 200 \times 10^9\ \text{N/m}^2 \left[(11.5 \times 10^{-6}/°\text{C} \times 40°\text{C}) - \frac{0.04\ \text{mm}}{200\ \text{mm}}\right]$$

$$= 200 \times 10^9\ \text{N/m}^2 \times 2.6 \times 10^{-4}$$

$$= 52 \times 10^6\ \text{N/m}^2 \quad \text{or} \quad 52\ \text{MN/m}^2$$

i.e. the compressive stress in the spacer at 60°C is $52\ \text{MN/m}^2$.

Example 3 The gauge length of an unstressed tensile-test specimen at a temperature of 20°C was 50 mm. If the tensile test is conducted when the temperature of the specimen is 80°C, estimate the gauge length for a tensile stress of $150\ \text{MN/m}^2$. For the material, $E = 200\ \text{GN/m}^2$ and $\alpha = 12 \times 10^{-6}/°\text{C}$.

Total strain = temperature strain + strain due to stress

i.e. $\quad \epsilon = \alpha\Delta\theta + \sigma/E$

where $\alpha = 12 \times 10^{-6}/°\text{C}$ $\quad \Delta\theta = 80°\text{C} - 20°\text{C} = 60°\text{C}$

$\sigma = 150\ \text{MN/m}^2 = 150 \times 10^6\ \text{N/m}^2$ and $E = 200 \times 10^9\ \text{N/m}^2$

$$\therefore \quad \epsilon = 12 \times 10^{-6}/°\text{C} \times 60°\text{C} + \frac{150 \times 10^6\ \text{N/m}^2}{200 \times 10^9\ \text{N/m}^2}$$

$$= (0.72 \times 10^{-3}) + (0.75 \times 10^{-3})$$

$$= 1.47 \times 10^{-3}$$

Increase in length, $x = \epsilon l$

$$= 1.47 \times 10^{-3} \times 50\ \text{mm}$$

$$= 0.074\ \text{mm}$$

$\therefore$ final gauge length $= 50\ \text{mm} + 0.074\ \text{mm}$

$$= 50.074\ \text{mm}$$

i.e. the final gauge length is 50.074 mm.

1.7 Effect of temperature change on composite bars

Consider a composite bar consisting of a copper core A placed inside a steel tube B, with the end faces of the core and the tube being firmly attached to end plates as shown in fig. 1.6(a). When the assembly is at a temperature θ_1, A and B are of equal length l and are assumed to be unstressed.

If one of the end plates is removed and the temperature of the assembly is increased to θ_2, A and B will expand freely to the positions shown in

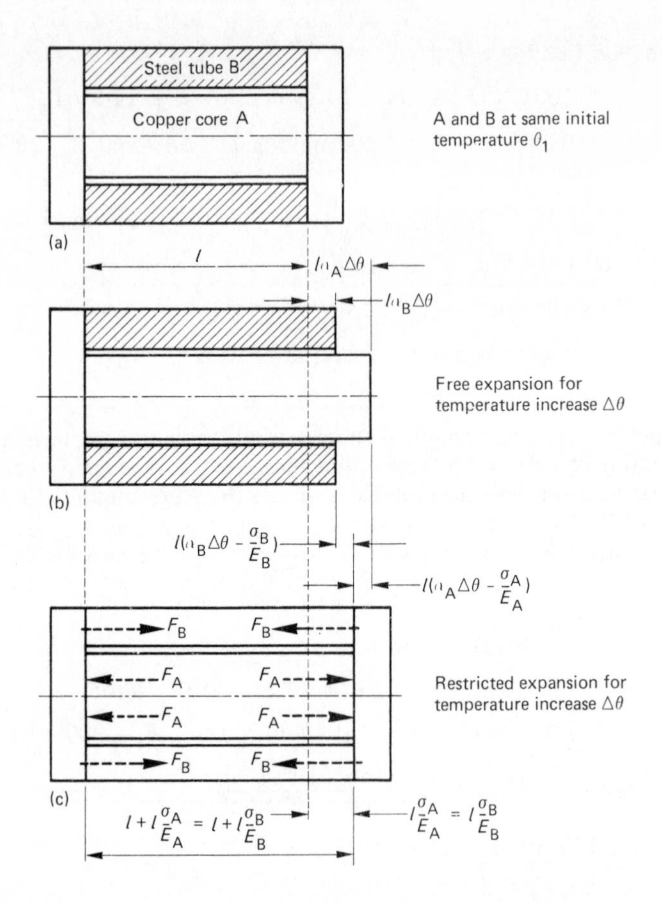

Fig. 1.6 Effect of temperature increase on a compound bar

fig. 1.6(b). Referring to fig. 1.6(b), the copper core A has expanded more than the steel tube B because the coefficient of linear expansion of copper is greater than that for steel.

If, with both end plates in position, the temperature of the assembly is again increased from θ_1 to θ_2, then, provided there is no distortion in the end plates, the copper and the steel will expand an *equal* amount as shown in fig. 1.6(c). Referring to fig. 1.6(c), the length of the copper core A is *reduced* and the length of the steel B is *increased* from the position shown in fig. 1.6(b); i.e. the copper core A is now being subjected to a *compressive* stress σ_A and the steel tube B to a *tensile* stress σ_B.

From section 1.6,

$$\text{total direct strain} = \text{sum of strains due to temperature change and stress}$$

or

$$\epsilon = \alpha\Delta\theta \pm \sigma/E$$

For the copper core A,

$$\epsilon_A = \alpha_A \Delta\theta - \sigma_A / E_A$$

For the steel tube B,

$$\epsilon_B = \alpha_B \Delta\theta + \sigma_B / E_B$$

where $\Delta\theta = \theta_2 - \theta_1$

At temperature θ_2,

length of A = length of B

i.e. total direct strain in A, ϵ_A = total direct strain in B, ϵ_B

or $$\alpha_A \Delta\theta - \sigma_A / E_A = \alpha_B \Delta\theta + \sigma_B / E_B$$

$$\therefore \quad (\alpha_A - \alpha_B)\,\Delta\theta = \frac{\sigma_A}{E_A} + \frac{\sigma_B}{E_B}$$

i.e. difference between the temperature strains = sum of the strains due to the stresses

which it is useful to remember.

Referring to fig. 1.6(c), the copper core A is exerting a *pushing* force F_A on the end plates, while the steel tube B is exerting a *pulling* force F_B. Since the assembly is in equilibrium,

force exerted by A = force exerted by B

or $$F_A = F_B$$

If A_A and A_B represent the cross-sectional areas of A and B respectively, then $F_A = \sigma_A A_A$ and $F_B = \sigma_B A_B$

$$\therefore \quad \sigma_A A_A = \sigma_B A_B$$

which it is useful to remember.

Example 1 If the diameter of the copper core in the composite bar shown in fig. 1.6(a) is 45 mm, and the internal and external diameters of the steel tube are 50 mm and 80 mm respectively, determine the stresses in the copper and the steel for a temperature increase of 60°C if both are assumed to be initially unstressed. For steel, $E = 200 \times 10^9\ \text{N/m}^2$ and $\alpha = 11.5 \times 10^{-6}/°\text{C}$. For copper, $E = 120 \times 10^9\ \text{N/m}^2$ and $\alpha = 16.5 \times 10^{-6}/°\text{C}$.

Let subscript A refer to the copper and subscript B refer to the steel.

$$\sigma_A A_A = \sigma_B A_B$$

$$\therefore \quad \sigma_A = \sigma_B (A_B / A_A)$$

where $A_A = (\pi/4) \times (45\ \text{mm})^2 = 1590\ \text{mm}^2$

and $A_B = (\pi/4) \times (80^2 - 50^2)\ \text{mm}^2 = 3063\ \text{mm}^2$

$$\therefore \quad \sigma_A = \frac{3063\ \text{mm}^2}{1590\ \text{mm}^2} \times \sigma_B = 1.93\,\sigma_B$$

$$(\alpha_A - \alpha_B)\,\Delta\theta = \sigma_A/E_A + \sigma_B/E_B$$
$$= 1.93\,\sigma_B/E_A + \sigma_B/E_B$$

i.e.
$$\sigma_B = \frac{(\alpha_A - \alpha_B)\,\Delta\theta}{1/E_B + 1.93/E_A}$$

where $\alpha_A = 16.5 \times 10^{-6}/°\text{C}$ $\quad \alpha_B = 11.5 \times 10^{-6}/°\text{C}$ $\quad \Delta\theta = 60°\text{C}$

$E_A = 120 \times 10^9\ \text{N/m}^2$ and $E_B = 200 \times 10^9\ \text{N/m}^2$

$$\therefore \quad \sigma_B = \frac{(16.5 - 11.5) \times 10^{-6}/°\text{C} \times 60°\text{C}}{1.93/(120 \times 10^9\ \text{N/m}^2) + 1/(200 \times 10^9\ \text{N/m}^2)}$$

$$= \frac{3 \times 10^{-4}}{(1.608 \times 10^{-11}) + (0.5 \times 10^{-11})}\ \text{N/m}^2$$

$$= 14.23 \times 10^6\ \text{N/m}^2 \quad \text{or} \quad 14.23\ \text{MN/m}^2$$

and $\sigma_A = 1.93\,\sigma_B$

$$= 1.93 \times 14.23\ \text{MN/m}^2$$
$$= 27.46\ \text{MN/m}^2$$

i.e. the stresses in the steel and copper after the temperature increase are $14.23\ \text{MN/m}^2$ and $27.46\ \text{MN/m}^2$ respectively.

Example 2 The assembly shown in fig. 1.7 consists of a brass cylinder clamped between flanges by a steel stud. The cylinder is 60 mm diameter x 45 mm bore and the stud is 12 mm diameter. At a temperature of 10°C, the tensile stress in the stud is $60\ \text{N/mm}^2$. What is the compressive stress in the cylinder material? If the temperature of the assembly is now increased to 40°C, calculate the total stresses in the cylinder and stud materials. Ignore the effect of temperature on the flanges. For steel, $E = 200\ \text{kN/mm}^2$ and $\alpha = 11.5 \times 10^{-6}/°\text{C}$. For brass, $E = 90\ \text{kN/mm}^2$ and $\alpha = 17 \times 10^{-6}/°\text{C}$.

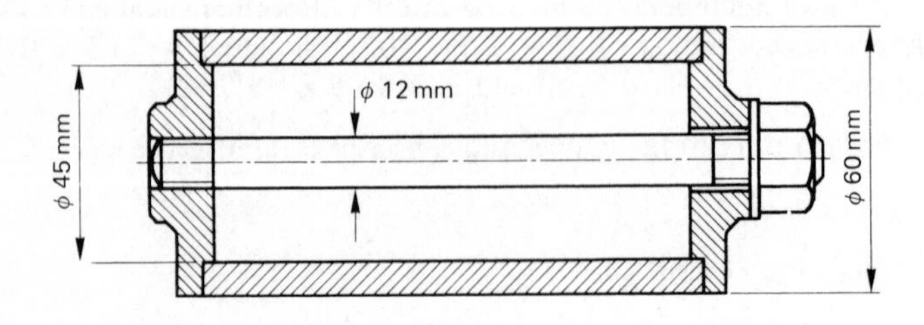

Fig. 1.7

Let subscript A refer to the brass and subscript B refer to the steel.
Referring to fig. 1.7,

force on cylinder = force on stud

i.e. $\sigma_A A_A = \sigma_B A_B$

or $\sigma_A = \sigma_B (A_B / A_A)$

where $A_A = (\pi/4) \times (60^2 - 45^2)\,\text{mm}^2 = 1237\,\text{mm}^2$

and $A_B = (\pi/4) \times (12\,\text{mm})^2 = 113.1\,\text{mm}^2$

$$\therefore \quad \sigma_A = \frac{113.1\,\text{mm}^2}{1237\,\text{mm}^2} \times \sigma_B$$

$$= 0.091\,\sigma_B$$

At 10°C,

$$\sigma_{B,10} = 60\,\text{N/mm}^2$$

$$\therefore \quad \sigma_{A,10} = 0.091 \times 60\,\text{N/mm}^2$$

$$= 5.46\,\text{N/mm}^2$$

i.e. at a temperature of 10°C, the compressive stress in the cylinder is $5.46\,\text{N/mm}^2$.

$$(\alpha_A - \alpha_B)\,\Delta\theta = \sigma_A/E_A + \sigma_B/E_B$$

$$= 0.091\,\sigma_B/E_A + \sigma_B/E_B$$

i.e. $$\sigma_B = \frac{(\alpha_A - \alpha_B)\,\Delta\theta}{0.091/E_A + 1/E_B}$$

where $\alpha_A = 17 \times 10^{-6}/°\text{C}$ $\qquad \alpha_B = 11.5 \times 10^{-6}/°\text{C}$

$\Delta\theta = 40°\text{C} - 10°\text{C} = 30°\text{C}$

$E_A = 90\,\text{kN/mm}^2 = 90 \times 10^3\,\text{N/mm}^2$

and $E_B = 200\,\text{kN/mm}^2 = 200 \times 10^3\,\text{N/mm}^2$

At 40°C,

$$\sigma_{B,40} = \frac{(17 - 11.5) \times 10^{-6}/°\text{C} \times 30°\text{C}}{0.091/(90 \times 10^3\,\text{N/mm}^2) + 1/(200 \times 10^3\,\text{N/mm}^2)}$$

$$= \frac{1.65 \times 10^{-4}}{(1.011 + 5) \times 10^{-6}}\,\text{N/mm}^2$$

$$= 27.45\,\text{N/mm}^2$$

i.e. the tensile stress in the stud is increased by $27.45\,\text{N/mm}^2$,

$\therefore$ total stress in the stud at 40°C $= 60\,\text{N/mm}^2 + 27.45\,\text{N/mm}^2$

$$= 87.45\,\text{N/mm}^2$$

and total stress in the brass at 40°C = $87.45\ \text{N/mm}^2 \times 0.091$

$= 7.96\ \text{N/mm}^2$

i.e. the total stresses in the cylinder and stud materials at a temperature of 40°C are $7.96\ \text{N/mm}^2$ and $87.5\ \text{N/mm}^2$ respectively.

1.8 Shear stress

Shear stress is defined as shear force F per unit cross-sectional area A resisting the shear,

i.e. $$\text{shear stress} = \frac{\text{shear force}}{\text{cross-sectional area resisting shear}}$$

The symbol used for shear stress is τ (*tau*),

i.e. $$\tau = \frac{F}{A}$$

which should be remembered.

The basic unit for shear stress is the newton per square metre (N/m^2).

Example 1 The simple riveted lap joint shown in fig. 1.8 contains six rivets, each 8 mm diameter, and is subjected to a shear force of 1.6 kN. Determine the shear stress in each rivet.

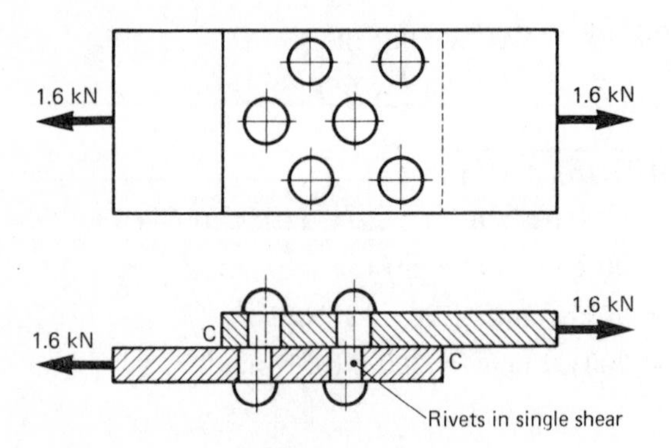

Fig. 1.8 Simple riveted lap joint

Referring to fig. 1.8, the joint will shear at the interface CC,

i.e. area resisting shear, A = total cross-sectional area of all the rivets

$= (\pi/4) \times (8\ \text{mm})^2 \times 6$ rivets

$= 301.6\ \text{mm}^2$

Shear stress $\tau = F/A$

where $F = 1.6\ \text{kN} = 1600\ \text{N}$

$$\therefore \qquad \tau = \frac{1600\ \text{N}}{301.6\ \text{mm}^2}$$

$$= 5.3\ \text{N/mm}^2$$

i.e. the shear stress in each rivet is $5.3\ \text{N/mm}^2$.

Note that the area resisting shear in a riveted joint is determined by the number of shear planes and the minimum number of rivets which must be removed to break the joint.

Example 2 The double-strap butt joint shown in fig. 1.9 contains eight rivets, each of diameter 10 mm. If the shear stress in the rivet material is not to exceed $20\ \text{N/mm}^2$, find the maximum shear force the joint can support.

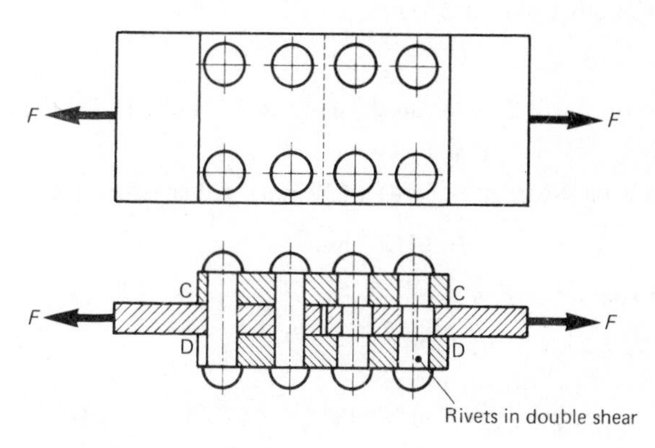

Fig. 1.9 Double-strap butt joint

Referring to fig. 1.9, the joint will shear at the interfaces CC and DD, i.e. the rivets are in *double* shear. Also, the joint will break if a minimum of *four* rivets are removed.

i.e. area resisting shear, $A = (\pi/4) \times (10\ \text{mm})^2 \times 4\ \text{rivets} \times 2\ \text{shear planes}$

$$= 628.32\ \text{mm}^2$$

Shear stress $\tau = F/A$

$$\therefore \qquad F = \tau A$$

where $\tau = 20\ \text{N/mm}^2$

$$\therefore \qquad F = 20\ \text{N/mm}^2 \times 628.32\ \text{mm}^2$$

$$= 12.566 \times 10^3\ \text{N} \quad \text{or} \quad 12.566\ \text{kN}$$

i.e. the maximum shear force is 12.566 kN.

Example 3 Figure 1.10 shows the coupling arrangement for a trailer. If the pin is 30 mm diameter, determine the shear stress in the pin when the axial pull is 8.5 kN.

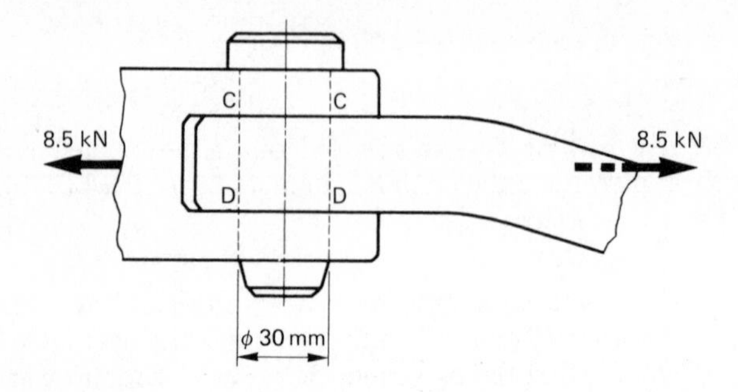

Fig. 1.10 Coupling pin for a trailer

Referring to fig. 1.10, the pin can shear across CC and DD, i.e. the pin is in double shear,

$$\therefore \quad \text{area resisting shear, } A = (\pi/4) \times (30\text{ mm})^2 \times 2 \text{ shear planes}$$

$$= 1413.7\text{ mm}^2$$

$$\text{Shear stress } \tau = F/A$$

where $F = 8.5\text{ kN} = 8500\text{ N}$

$$\therefore \quad \tau = \frac{8500\text{ N}}{1413.7\text{ mm}^2} = 6\text{ N/mm}^2$$

i.e. the shear stress in the pin material is 6 N/mm^2.

Example 4 In a blanking operation, the force on the punch was found to be 20 kN. If the diameter of the punch was 35 mm and the material was 3.6 mm thick, determine the shear stress in the material.

Figure 1.11 shows the blanked portion.

$$\begin{matrix}\text{Area resisting} \\ \text{blanking force}\end{matrix} = \begin{matrix}\text{circumference} \\ \text{of blank}\end{matrix} \times \begin{matrix}\text{material} \\ \text{thickness}\end{matrix}$$

$$= \pi \times 35\text{ mm} \times 3.6\text{ mm}$$

$$= 395.8\text{ mm}^2$$

$$\text{Shear stress } \tau = F/A$$

$$\therefore \quad \tau = \frac{20 \times 10^3\text{ N}}{395.8\text{ mm}^2} = 50.5\text{ N/mm}^2$$

i.e. the shear stress was 50.5 N/mm^2.

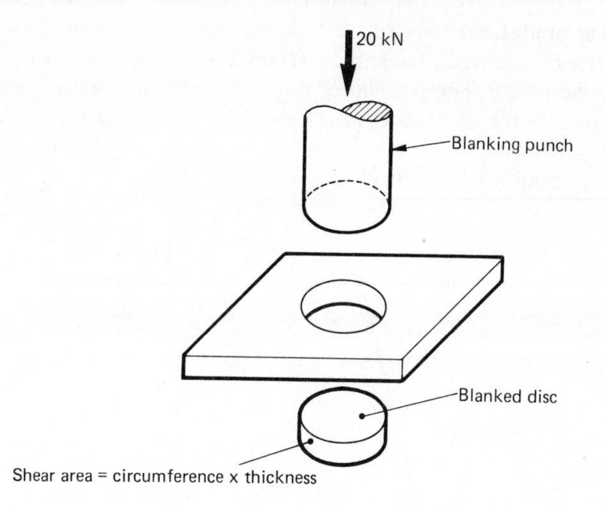

Fig. 1.11

1.9 Shear strain

Shear strain is defined as the angular distortion a plane normal to the shear force suffers as a result of the shear force. Figure 1.12 shows a block ABCD subjected to a shear force F. Before loading, the planes AB and CD are normal (i.e. at 90°) to F. After loading, the planes are distorted through angle γ (*gamma*) to AB′ and DC′ respectively. The angle γ, which is measured in radians, is the *shear strain*.

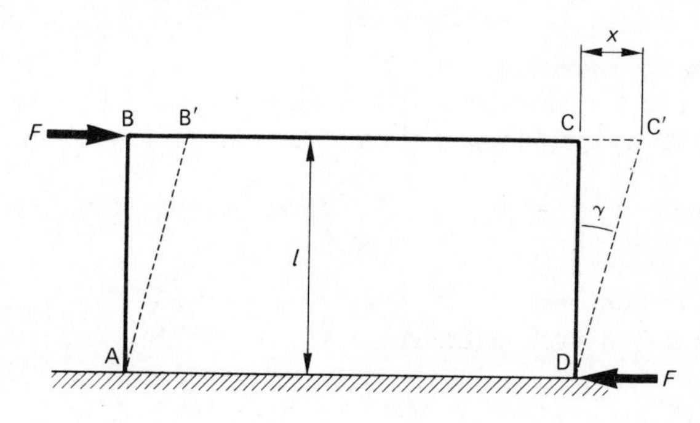

Fig. 1.12 Shear strain

Referring to fig. 1.12,

$$\text{shear strain } \gamma = \frac{x}{l}$$

which it is useful to remember.

1.10 Shear modulus

If shear stress is plotted against shear strain, a graph similar to that shown in fig. 1.13 will result. The gradient or slope of the linear portion of the graph OA is known as the *shear modulus* or *modulus of rigidity* for the material,

$$\text{i.e.} \quad \text{shear modulus} \equiv \text{gradient of OA} = \frac{\text{shear stress}}{\text{shear strain}}$$

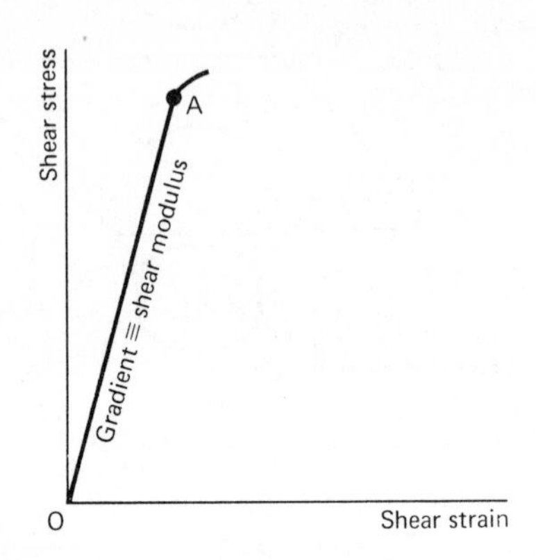

Fig. 1.13 Graph of shear stress against shear strain for a material

The symbol used for the shear modulus is G,

$$\text{i.e.} \quad G = \frac{\tau}{\gamma}$$

which should be remembered.

The basic unit for shear modulus is the newton per square metre (N/m^2). Table 1.2 shows typical values of the shear modulus for various materials.

Material	*Shear modulus* (GN/m^2)
0.25% carbon steel	82.2
0.75% carbon steel	81.1
0.75% carbon steel, hardened	77.8
Cast iron	60
Copper	48.3
70/30 brass	48.3
Aluminium	26.1

Table 1.2 Typical values of the shear modulus

Example 1 The maximum shear stress in a shaft was found to be 45 MN/m^2. If G for the shaft material was 79 GN/m^2, determine the shear strain.

$$G = \tau/\gamma$$

$$\therefore \quad \gamma = \tau/G$$

where $\tau = 45\ \text{MN/m}^2 = 45 \times 10^6\ \text{N/m}^2$
and $G = 79\ \text{GN/m}^2 = 79 \times 10^9\ \text{N/m}^2$

$$\therefore \quad \gamma = \frac{45 \times 10^6\ \text{N/m}^2}{79 \times 10^9\ \text{N/m}^2} = 0.57 \times 10^{-3}\ \text{rad}$$

i.e. the maximum shear strain was 0.57×10^{-3} radians.

Example 2 The apparatus shown in fig. 1.14 was used to determine the shear modulus of rubber. When the mean load was 100 N, the deflection was 1.2 mm. Calculate the shear modulus of the rubber.

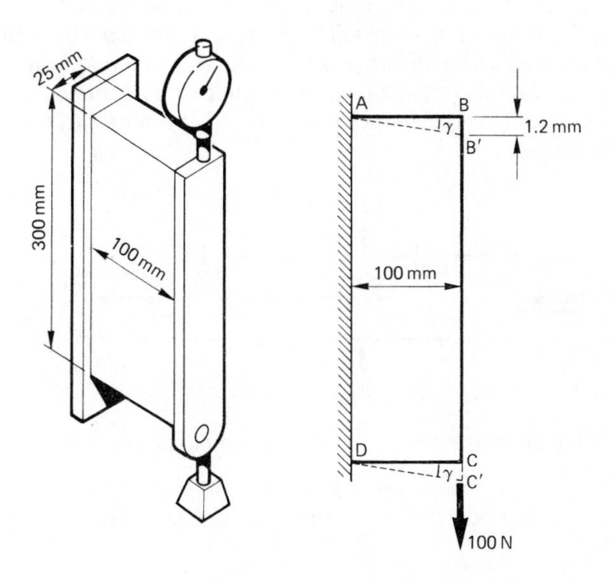

Fig. 1.14 Apparatus to determine shear modulus of rubber

Referring to fig. 1.14,

$$\text{area resisting shear} = 300\ \text{mm} \times 25\ \text{mm} = 7500\ \text{mm}^2$$

$$\text{Shear stress } \tau = F/A$$

where $F = 100\ \text{N}$

$$\therefore \quad \tau = \frac{100\ \text{N}}{7500\ \text{mm}^2} = 0.0133\ \text{N/mm}^2$$

After loading, the faces AB and DC are distorted through angle γ to AB′ and DC′ respectively.

$$\therefore \quad \gamma = \frac{1.2\ \text{mm}}{100\ \text{mm}} = 0.012\ \text{rad}$$

$$\text{Shear modulus } G = \tau/\gamma$$

$$= \frac{0.0133\ \text{N/mm}^2}{0.012\ \text{rad}}$$

$$= 1.11\ \text{N/mm}^2 \quad \text{or} \quad 1.11\ \text{MN/m}^2$$

i.e. the shear modulus of the rubber is 1.11 MN/m^2.

1.11 Poisson's ratio

As the circular specimen shown in fig. 1.15 is stretched, its diameter will be reduced, i.e. a compressive strain will be induced in the material at right angles to the line of action of the tensile force. This strain, known as the *lateral* strain, is directly proportional to the direct or *longitudinal* strain produced by the force; i.e. lateral strain is directly proportional to longitudinal strain.

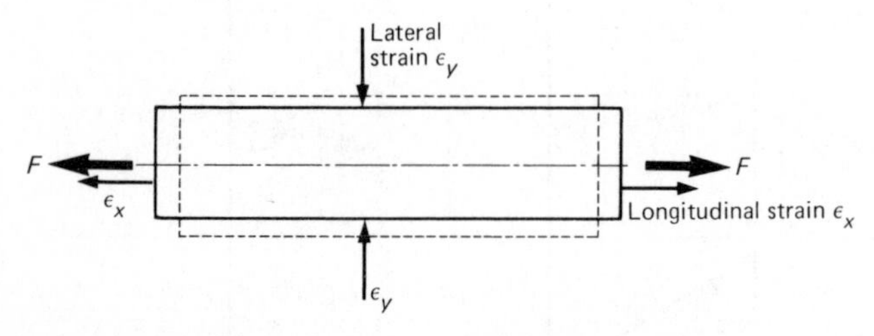

Fig. 1.15 Lateral compression due to longitudinal force

Let ϵ_x = longitudinal strain and ϵ_y = lateral strain

then $\epsilon_y \propto \epsilon_x$

or $\epsilon_y = \nu\, \epsilon_x$

The constant ν (*nu*) is known as Poisson's ratio,

$$\text{i.e.} \quad \text{Poisson's ratio } \nu = \frac{\epsilon_y}{\epsilon_x} = \frac{\text{lateral strain}}{\text{longitudinal strain}}$$

which should be remembered.

Since it is a ratio of like quantities, Poisson's ratio has no units. For metals, Poisson's ratio can vary between 0.25 and 0.33.

Example A tensile-test specimen of diameter 10 mm is subjected to an axial force of 15 kN. If E for the material is 207 kN/mm^2 and Poisson's ratio is 0.29, determine the lateral strain and the change in diameter.

Longitudinal strain $\epsilon_x = \sigma/E = F/(AE)$

where $F = 15\,\text{kN} = 15 \times 10^3\,\text{N}$ $\quad A = (\pi/4) \times (10\,\text{mm})^2 = 78.54\,\text{mm}^2$

and $E = 207\,\text{kN/mm}^2 = 207 \times 10^3\,\text{N/mm}^2$

$$\therefore \quad \epsilon_x = \frac{15 \times 10^3\,\text{N}}{78.54\,\text{mm}^2 \times 207 \times 10^3\,\text{N/mm}^2}$$

$$= 9.23 \times 10^{-4}$$

Lateral strain $= \nu \times$ longitudinal strain

or $\quad \epsilon_y = \nu\,\epsilon_x$

where $\quad \nu = 0.29$

$$\therefore \quad \epsilon_y = 0.29 \times 9.23 \times 10^{-4}$$

$$= 2.68 \times 10^{-4}$$

i.e. the lateral strain is 2.68×10^{-4}.

Change in diameter $=$ original diameter $\times$ lateral strain

$= 10\,\text{mm} \times 2.68 \times 10^{-4}$

$= 2.68 \times 10^{-3}\,\text{mm}$

i.e. the change in diameter is 2.68×10^{-3} mm.

1.12 Two-dimensional stress

Consider the stressed element shown in fig. 1.16(a). The strains induced in the element by σ_x and σ_y each acting alone are shown in figs 1.16(b) and 1.16(c) respectively. Figure 1.16(d) illustrates the total or resultant strains in the directions of σ_x and σ_y.

Referring to fig. 1.16(d), the resultant strain in the direction of σ_x is given by

$$\epsilon_x = \frac{\sigma_x}{E} - \nu\,\frac{\sigma_y}{E}$$

and the resultant strain in the direction of σ_y is given by

$$\epsilon_y = \frac{\sigma_y}{E} - \nu\,\frac{\sigma_x}{E}$$

which should be remembered.

Example 1 An element is subjected to two mutually perpendicular tensile stresses of magnitude $30\,\text{MN/m}^2$ and $20\,\text{MN/m}^2$ respectively. If $E = 200\,\text{GN/m}^2$ and Poisson's ratio is 0.3, determine the resultant strains in the directions of the stresses.

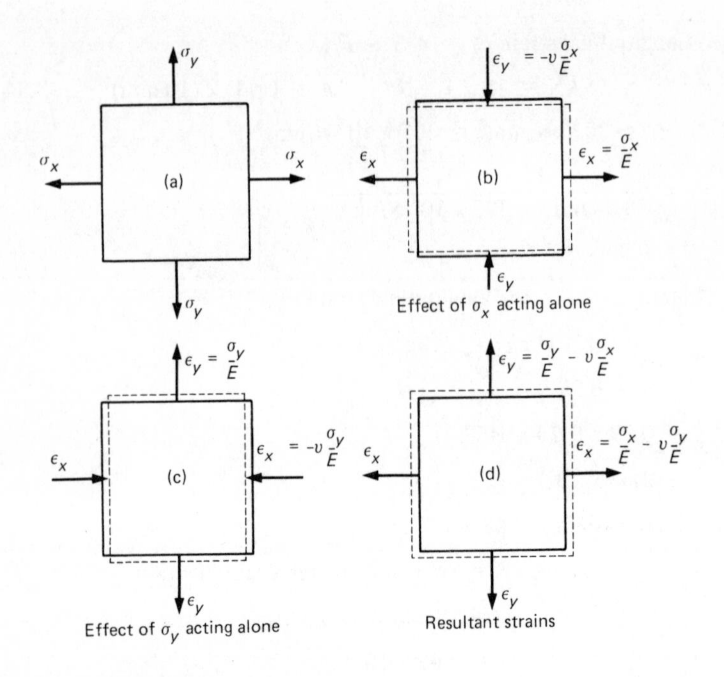

Fig. 1.16 Two-dimensional stress

Let $\sigma_x = 30\ \text{MN/m}^2 = 30 \times 10^6\ \text{N/m}^2$

and $\sigma_y = 20\ \text{MN/m}^2 = 20 \times 10^6\ \text{N/m}^2$

then $\epsilon_x = \sigma_x/E - \nu(\sigma_y/E)$

where $E = 200\ \text{GN/m}^2 = 200 \times 10^9\ \text{N/m}^2$ and $\nu = 0.3$

$$\therefore \quad \epsilon_x = \left(\frac{30 \times 10^6\ \text{N/m}^2}{200 \times 10^9\ \text{N/m}^2}\right) - \left(0.3 \times \frac{20 \times 10^6\ \text{N/m}^2}{200 \times 10^9\ \text{N/m}^2}\right)$$

$$= (1.5 \times 10^{-4}) - (0.3 \times 10^{-4})$$

$$= 1.2 \times 10^{-4}$$

i.e. the strain in the direction of the $30\ \text{MN/m}^2$ stress is 1.2×10^{-4}.

Also, $\epsilon_y = \sigma_y/E - \nu(\sigma_x/E)$

$$= \left(\frac{20 \times 10^6\ \text{N/m}^2}{200 \times 10^9\ \text{N/m}^2}\right) - \left(0.3 \times \frac{30 \times 10^6\ \text{N/m}^2}{200 \times 10^9\ \text{N/m}^2}\right)$$

$$= (1 \times 10^{-4}) - (0.45 \times 10^{-4})$$

$$= 0.55 \times 10^{-4}$$

i.e. total strain in direction of the $20\ \text{MN/m}^2$ stress is 0.55×10^{-4}.

Example 2 Determine the total strains in the directions of the stresses shown in fig. 1.17, if $E = 120\ \text{GN/m}^2$ and $\nu = 0.28$.

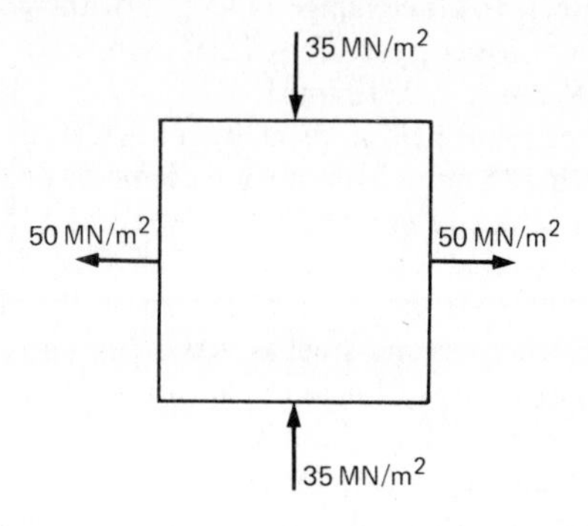

Fig. 1.17

Let $\sigma_x = 50\ \text{MN/m}^2 = 50 \times 10^6\ \text{N/m}^2$
and $\sigma_y = -(35\ \text{MN/m}^2) = -(35 \times 10^6)\ \text{N/m}^2$

(σ_y is negative since the stress is compressive and of *opposite* sign to the tensile stress.)

then $\epsilon_x = \sigma_x/E - \nu(\sigma_y/E)$

where $E = 120\ \text{GN/m}^2 = 120 \times 10^9\ \text{N/m}^2$ and $\nu = 0.28$

$$\therefore \quad \epsilon_x = \left(\frac{50 \times 10^6\ \text{N/m}^2}{120 \times 10^9\ \text{N/m}^2}\right) - \left(0.28 \times \frac{-(35 \times 10^6)\ \text{N/m}^2}{120 \times 10^9\ \text{N/m}^2}\right)$$

$$= (4.17 \times 10^{-4}) + (0.82 \times 10^{-4})$$

$$= 4.99 \times 10^{-4}$$

i.e. the strain in the direction of the $50\ \text{MN/m}^2$ stress is 4.99×10^{-4}.

Also, $\epsilon_y = \sigma_y/E - \nu(\sigma_x/E)$

$$= \left(\frac{-(35 \times 10^6)\ \text{N/m}^2}{120 \times 10^9\ \text{N/m}^2}\right) - \left(0.28 \times \frac{50 \times 10^6\ \text{N/m}^2}{120 \times 10^9\ \text{N/m}^2}\right)$$

$$= (-2.92 \times 10^{-4}) - (1.17 \times 10^{-4})$$

$$= -4.09 \times 10^{-4}$$

(the minus sign indicating that the strain is compressive)
i.e. the total strain in the direction of the $35\ \text{MN/m}^2$ stress is 4.09×10^{-4} and it is compressive.

Exercises on chapter 1

1 A composite bar is 350 mm long and consists of a copper tube butt-welded to the end face of a steel bar. The tube dimensions are 30 mm diameter

x 20 mm bore x 200 mm long, and the steel bar is 30 mm diameter and 150 mm long. If the bar supports an axial force of 4 kN, determine (a) the stresses in the copper and the steel, (b) the change in length of the composite bar. For steel, $E = 200 \text{ GN/m}^2$; for copper, $E = 120 \text{ GN/m}^2$.
[10.2 N/mm²; 5.66 N/mm²; 0.0212 mm]

2 A steel oil-filter casing is held in place by a brass stud, 10 mm diameter and of effective length 125 mm. The outside diameter of the casing is 55 mm, the material is 2.5 mm thick, and its effective length is 125 mm. If the clamping force on the stud at a temperature of 15°C is 1500 N, at what oil temperature will the filter just begin to leak? Ignore the effect of the filter end plate and take E for brass and steel as 90 GN/m^2 and 200 GN/m^2 respectively. The coefficients of linear expansion for brass and steel are $16 \times 10^{-6}/°\text{C}$ and $12 \times 10^{-6}/°\text{C}$. [72.6°C]

3 A concrete pillar, 250 mm diameter and 3 m long, is reinforced by steel rods, each 10 mm diameter and 3 m long. If the maximum stress in the concrete is not to exceed 5 N/mm^2, calculate the minimum number of steel rods required in order that the reinforced pillar can support an axial load of 600 kN. How much will the pillar shorten under this load? For steel, $E = 200 \text{ kN/mm}^2$; for concrete, $E = 10 \text{ kN/mm}^2$. [48; 1.5 mm]

4 Two steel plates are riveted together using ten 8 mm diameter rivets. If the shear stress in the rivet material is limited to 5 N/mm^2, determine the maximum shearing force which the joint can support. [2513 N]

5 The clevis pin in a steel sling is 30 mm diameter and is in double shear. What is the maximum load the pin can support if the shear stress is limited to 10 N/mm^2? [14 137 N]

6 The shear stress in a shaft transmitting power is 2.6 N/mm^2. If G for the shaft material is 79 GN/m^2, determine the shear strain. [3.3×10^{-5} rad]

7 At a point in a stressed material, the shear strain was found to be 4.8×10^{-4} rad. If the shear modulus for the material is 60 kN/mm^2, what was the shear stress? [28.8 N/mm²]

8 In an experiment to determine the shear modulus of steel, the following observations were made:

Shear stress (N/mm²)	5	10	15	20	25
Shear strain (rad x 10^{-5})	6.3	12.9	19.2	25.7	32.0

Plot a graph of shear stress against shear strain, and hence determine the shear modulus for the steel. [78 GN/m²]

9 A composite bar consists of a brass strip, 20 mm wide and 4 mm thick, sandwiched between two steel strips each 20 mm wide and 3 mm thick, the three strips being held together by two rivets each 6 mm diameter and 50 mm apart. The shear stress at failure in the rivet material is 8.49 N/mm^2. If the assembly was made at a temperature of 16°C and is now heated, at what temperature will the rivets fail? What will be the distance between the rivets at the instant of failure? Ignore the effect of temperature on the rivet material. For steel, $E = 200 \text{ kN/mm}^2$ and $\alpha = 11.5 \times 10^{-6}/°\text{C}$.
For brass, $E = 90 \text{ kN/mm}^2$ and $\alpha = 16 \times 10^{-6}/°\text{C}$. [25.6°C; 50.006 mm]

10 A steel rope of diameter 25 mm and length 4 m is subjected to a tensile force of 30 kN. If the modulus of elasticity and Poisson's ratio for the material are 200 GN/m^2 and 0.28 respectively, determine the final length and diameter of the rope. [4.0012 m; 24.998 mm]

11 The clearance on diameter between a bearing and a shaft of diameter 200 mm is 0.04 mm. If the modulus of elasticity and Poisson's ratio for the shaft material are 207 GN/m^2 and 0.29 respectively, determine the axial force which must be applied to the shaft to reduce the diametral clearance to 0.01 mm. [3.36 MN]

12 A material is subjected to two mutually perpendicular tensile stresses of magnitudes 40 N/mm^2 and 60 N/mm^2 respectively. If Poisson's ratio is 0.3 and the modulus of elasticity is 150 kN/mm^2, determine the total strains in the directions of the stresses. [1.47×10^{-4}; 3.2×10^{-4}]

13 At a point in a material, there is a compressive stress of 60 N/mm^2 and perpendicular to this stress a tensile strain of 2×10^{-4}. Determine (a) the magnitude and sense of the stress in the plane of the tensile strain, (b) the magnitude and sense of the strain in the direction of the compressive stress. For the material, $E = 207$ kN/mm^2 and $\nu = 0.28$.
[25 N/mm^2 (tensile); 3.23×10^{-4} (compressive)]

14 A short concrete column has a square cross-section of side 400 mm and is reinforced by four steel bars, each 20 mm diameter. If the ratio E_{steel} to $E_{concrete}$ is 15:1, calculate the compressive stress in the concrete when the column supports an axial load of 1.4 MN. [7.88 MN/m^2]

15 A double-strap butt joint contains a total of twelve rivets, each 5 mm diameter. Calculate the shear stress in each rivet when the joint is subjected to a shearing force of 1 kN. [8.49 N/mm^2]

16 Define Poisson's ratio.

At a point in a material, there is a tensile strain of 8×10^{-4} and perpendicular to this strain a compressive strain of 3×10^{-4}. If Poisson's ratio is 0.3 and the modulus of elasticity is 150 GN/m^2, calculate the stresses in the direction of the strains.
[117 MN/m^2 (tensile); 9.89 MN/m^2 (compressive)]

17 A steel strip, 50 mm x 10 mm x 200 mm long, is subjected to an axial tensile force of 125 kN. If Poisson's ratio = 0.3 and $E = 200$ GN/m^2, determine the change in volume. [49.92 mm^3 increase]

18 A steel tube having an outside diameter of 75 mm is 'shrink-fitted' on to a copper cylinder of 65 mm outside diameter and 40 mm bore. After assembly, the ends of the composite tube are machined to give an overall length of 150 mm. If the assembly is subjected to a temperature increase of 50°C, determine the shear stress at the copper–steel interface, assuming that the expansion of the copper and of the steel are the same. For the steel, $E = 200$ GN/m^2 and $\alpha = 11.5 \times 10^{-6}$/°C. For the copper, $E = 120$ GN/m^2 and $\alpha = 16.5 \times 10^{-6}$/°C. [0.95 N/$mm^2$]

2 Beams and bending

2.1 Introduction

A beam is part of a structure which supports forces perpendicular to its length, such forces being known as *transverse* forces. When transverse forces are applied, the beam will *bend*.

Figure 2.1 shows a simply supported beam with a transverse force F at mid-span. The beam bends under the force, causing the fibres in the plane of aa to be compressed and the fibres in the plane of bb to be stretched, setting up associated compressive and tensile stresses, called *bending stresses*, in the beam material. At a plane in the beam cross-section known as the *neutral plane* or *neutral axis*, there will be no change in the length of the fibres and thus no bending stress, i.e. the neutral axis is a plane of zero bending stress. It will be shown in section 2.4 that the neutral axis of a beam cross-section passes through the centroid of the section.

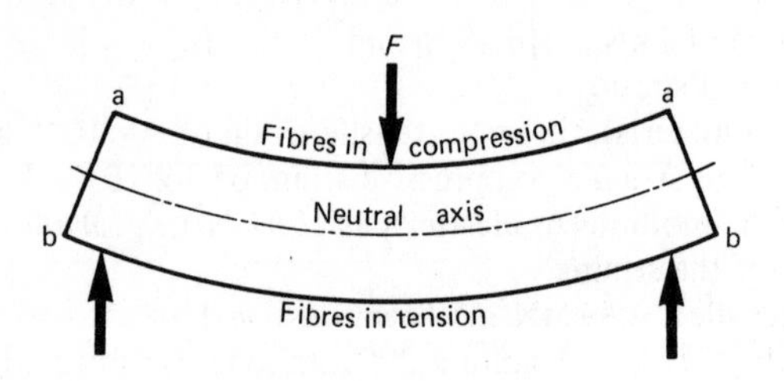

Fig. 2.1 Effect of transverse force on a beam

The magnitude of the bending stresses in a beam material will depend upon

a) the magnitude and position of the transverse forces relative to the beam supports – these control the *bending moment* at any section in the beam (see section 2.2);
b) the shape and size of the beam cross-section. The beam cross-section can be *any* shape; however, only sections which are *symmetrical* about the axis of loading will be considered here. Figure 2.2 illustrates a beam cross-section which may be both symmetrically and asymmetrically loaded.

2.2 Bending moment

The bending moment at any section in a beam which is in equilibrium is defined as the algebraic sum of the moments of all the forces acting *either* to

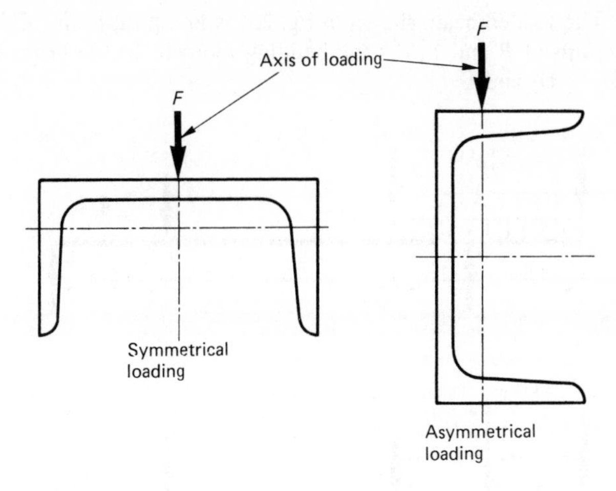

Fig. 2.2 Symmetrical and asymmetrical loading

the *left* of the section *or* to the *right* of the section being considered (*not both together*),

i.e. bending moment at any section $= \sum$ of the moments of all the forces acting to the *left* of the section

$= \sum$ of the moments of all the forces acting to the *right* of the section

The symbol used for bending moment is M.

When calculating bending moments, the sign convention shown in fig. 2.3 is normally used, i.e. forces which act upwards are *positive* forces and the resulting bending moment is also *positive*. Forces which act downwards are negative forces and the resulting bending moment is also *negative*. Positive bending causes the beam to 'sag', while negative bending causes the beam to 'hog' as shown.

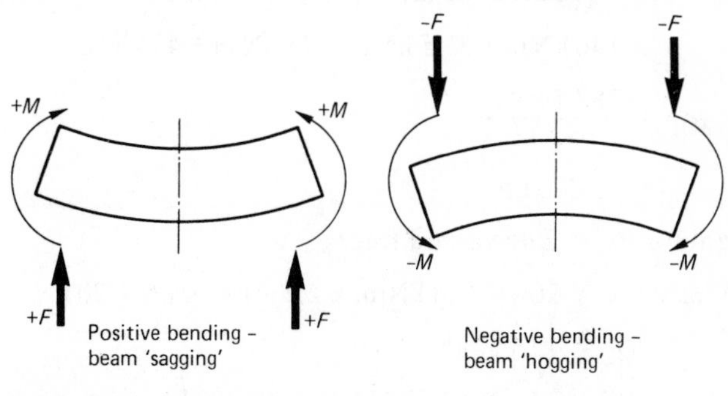

Fig. 2.3 Bending-moment sign convention

Example The loaded beam shown in fig. 2.4 is in equilibrium. Calculate (a) the reactions at B and E, (b) the bending moment in the beam at the points A, B, C, D, and E.

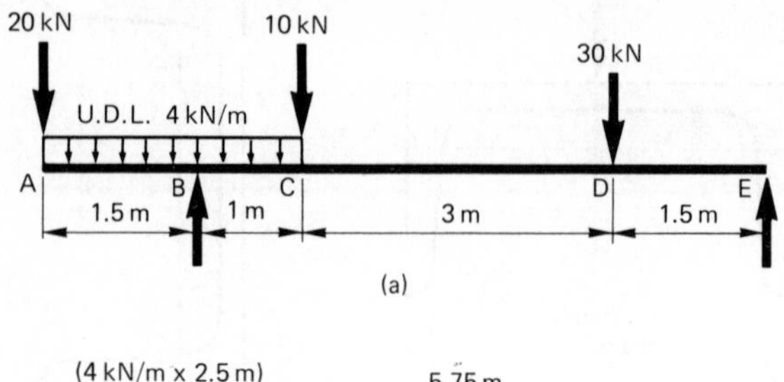

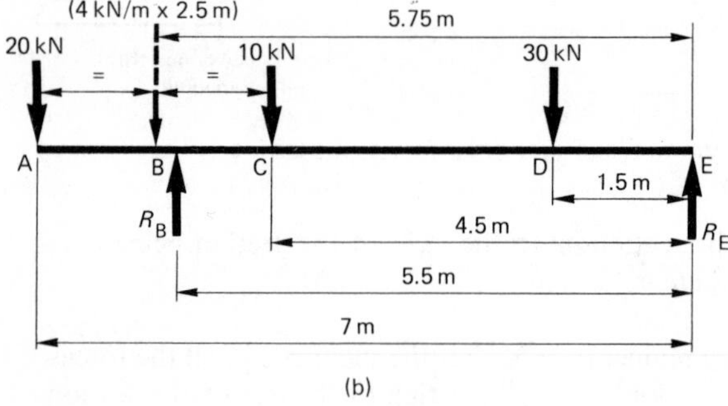

Fig. 2.4

a) To calculate the reactions at B and E, it is necessary first to replace the uniformly distributed load between A and C with a point load as shown dotted in fig. 2.4(b).

Taking moments about E,

$$\Sigma\text{clockwise moments} = \Sigma\text{anticlockwise moments}$$

$$\therefore \quad R_B \times 5.5\,\text{m} = [20\,\text{kN} \times 7\,\text{m}] + [(4\,\text{kN/m} \times 2.5\,\text{m}) \times 5.75\,\text{m}] + [10\,\text{kN} \times 4.5\,\text{m}] + [30\,\text{kN} \times 1.5\,\text{m}]$$

$$= 140\,\text{kN m} + 57.5\,\text{kN m} + 45\,\text{kN m} + 45\,\text{kN m}$$

$$\therefore \quad R_B = \frac{287.5\,\text{kN m}}{5.5\,\text{m}}$$

$$= 52.273\,\text{kN}$$

$$\Sigma\text{upward forces} = \Sigma\text{downward forces}$$

$$\therefore \quad R_E + 52.273\,\text{kN} = 20\,\text{kN} + (4\,\text{kN/m} \times 2.5\,\text{m}) + 10\,\text{kN} + 30\,\text{kN}$$

$$\therefore \quad R_E = 17.727\,\text{kN}$$

i.e. the reactions at B and E are 52.273 kN and 17.727 kN respectively.

b) Let the bending moments at A, B, C, D, and E be M_A, M_B, M_C, M_D, and M_E respectively,

then $M_A = \Sigma$ moments to *left* of A = 0

$M_B = \Sigma$ moments to *left* of B

$= [-20\,\text{kN} \times 1.5\,\text{m}] - [(4\,\text{kN/m} \times 1.5\,\text{m}) \times 0.75\,\text{m}]$

$= -34.5\,\text{kN m}$

i.e. at B the beam is 'hogging'.

$M_C = \Sigma$ moments to *left* of C

$= [-20\,\text{kN} \times 2.5\,\text{m}] - [(4\,\text{kN/m} \times 2.5\,\text{m}) \times 1.25\,\text{m}]$

$+ [52.273\,\text{kN} \times 1\,\text{m}]$

$= -10.23\,\text{kN m}$

i.e. at C the beam is 'hogging'.

Alternatively,

$M_C = \Sigma$ moments to *right* of C

$= [-30\,\text{kN} \times 3\,\text{m}] + [17.727\,\text{kN} \times 4.5\,\text{m}]$

$= -10.23\,\text{kN m}$

Notice that this calculation is slightly less laborious than the previous one.

$M_D = \Sigma$ moments to *left* of D

$= [-20\,\text{kN} \times 5.5\,\text{m}] - [(4\,\text{kN/m} \times 2.5\,\text{m}) \times 4.25\,\text{m}]$

$+ [52.273\,\text{kN} \times 4\,\text{m}] - [10\,\text{kN} \times 3\,\text{m}]$

$= 26.59\,\text{kN m}$

i.e. at D, the beam is 'sagging'.

Alternatively,

$M_D = \Sigma$ moments to the *right* of D

$= 17.727\,\text{kN} \times 1.5\,\text{m}$

$= 26.59\,\text{kN m}$

Notice that this calculation is *much* less laborious than the previous one!

$M_E = \Sigma$ moments to *right* of E = 0

i.e. the bending moments at A, B, C, D, and E are respectively 0, –34.5 kN m, –10.23 kN m, 26.59 kN m, and 0.

Notice, from the above example, that it is permissible to calculate the bending moments to the *left* or to the *right* of the section. The choice of direction will depend upon the number of forces involved. The bending-moment sign is important since this indicates the sense of the bending stresses in the upper and lower fibres of the beam cross-section, i.e. a negative bending moment indicates a tensile stress in the upper fibres and a compressive stress in the lower fibres, while a positive bending moment produces the opposite effect.

2.3 Distribution of bending stress

If the simply supported beam in fig. 2.5(a) is loaded as shown, it will bend to the shape of fig. 2.5(b), i.e. between the supports the bent shape of the beam is a circular arc of mean radius R. An enlarged view of a small portion of the beam between the supports is shown in figs 2.5(c) and (d).

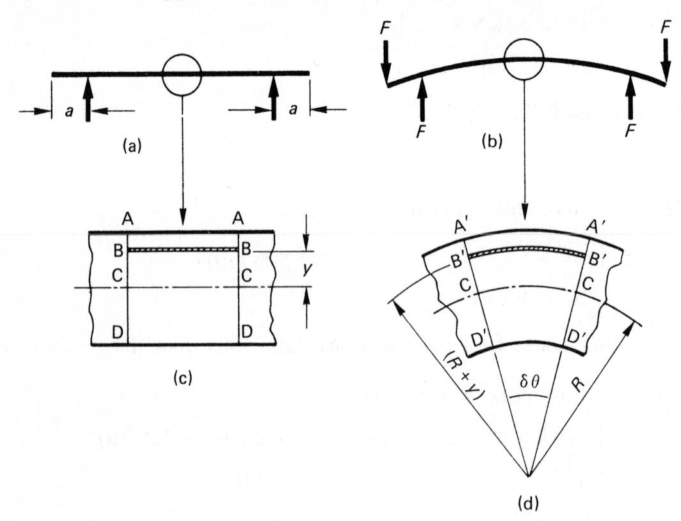

Fig. 2.5 Simply supported beam

Before bending (fig. 2.5(c)),

$$AA = BB = CC = DD$$

Consider the planes BB and CC after bending (fig. 2.5(d)). After bending, CC remains unchanged,

i.e. $\quad CC = R\delta\theta$

and BB increases to $\quad B'B' = (R+y)\,\delta\theta$

$\therefore\quad$ increase in length of BB $= B'B' - BB$

$$= (R+y)\,\delta\theta - R\delta\theta \quad \text{(since } BB = CC = R\delta\theta\text{)}$$

$$= y\delta\theta$$

Because the length of BB has been increased, a *strain* is induced in the beam material. From section 1.3,

$$\text{strain} = \frac{\text{change in dimension}}{\text{original dimension}}$$

$$= \frac{y\delta\theta}{R\delta\theta} = \frac{y}{R}$$

From section 1.4,

$$\text{modulus of elasticity } E = \frac{\text{stress } (\sigma)}{\text{strain } (\epsilon)}$$

$$\therefore \quad \text{strain} = \frac{\sigma}{E} = \frac{y}{R}$$

or $$\sigma = \frac{E}{R} y$$

Thus, *the bending stress at any plane in the beam section is directly proportional to the distance* y *from the neutral axis,* which it is useful to remember.

Figure 2.6(a) shows the distribution of bending stress across the beam cross-section of fig. 2.6(b). Above the neutral axis (n.a.), the tensile stress increases from zero (at the neutral axis) to σ_t. Below the neutral axis, the compressive stress increases from zero (at the neutral axis) to σ_c.

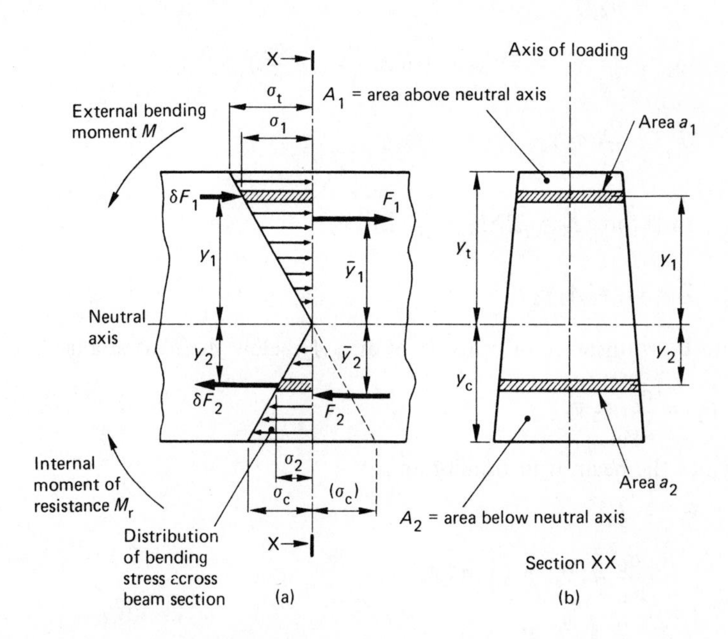

Fig. 2.6 Distribution of bending stress

2.4 Position of the neutral axis

Figure 2.6(a) shows the distribution of bending stress across the section of the beam in fig. 2.6(b) due to the external bending moment M. Consider the force δF_1 acting on the area a_1 at distance y_1 from the neutral axis,

i.e. $\delta F_1 = \sigma_1 a_1$

From fig. 2.6(a),

$$\sigma_1 = \frac{\sigma_t}{y_t} y_1$$

$$\therefore \quad \delta F_1 = \frac{\sigma_t}{y_t} a_1 y_1$$

and total force $F_1 = \Sigma \delta F_1 = \dfrac{\sigma_t}{y_t} \Sigma a_1 y_1$

But $\Sigma a_1 y_1 = A_1 \bar{y}_1$

where $\bar{y}_1$ = distance of centroid of area A_1 above the neutral axis

$$\therefore \quad F_1 = \frac{\sigma_t}{y_t} A_1 \bar{y}_1$$

Also, the force acting on area a_2 at distance y_2 below the neutral axis is given by

$$\delta F_2 = \sigma_2 a_2$$

But $\sigma_2 = \dfrac{\sigma_c}{y_c} y_2 = \dfrac{\sigma_t}{y_t} y_2$ (since $\dfrac{\sigma_t}{y_t} = \dfrac{\sigma_c}{y_c}$)

$$\therefore \quad \delta F_2 = \frac{\sigma_t}{y_t} a_2 y_2$$

and total force $F_2 = \Sigma \delta F_2 = \dfrac{\sigma_t}{y_t} \Sigma a_2 y_2$

But $\Sigma a_2 y_2 = A_2 \bar{y}_2$

where $\bar{y}_2$ = distance of centroid of area A_2 below the neutral axis

$$\therefore \quad F_2 = \frac{\sigma_t}{y_t} A_2 \bar{y}_2$$

Since the beam is in equilibrium,

$$F_1 = F_2$$

i.e. $$\frac{\sigma_t}{y_t} A_1 \bar{y}_1 = \frac{\sigma_t}{y_t} A_2 \bar{y}_2$$

$$\therefore \quad A_1 \bar{y}_1 = A_2 \bar{y}_2$$

or $\quad A_1\bar{y}_1 - A_2\bar{y}_2 = 0$

i.e. the algebraic sum of the first moments of area about the neutral axis is zero. By definition, for the algebraic sum of the first moments of area about a point in a plane area to be zero, that point must be the centroid of the area; thus, *the neutral axis must pass through the centroid of the section,* which should be remembered.

2.5 Relationship between bending stress and external bending moment

Referring to fig. 2.6(a), the forces δF_1 and δF_2 are exerting about the neutral axis a *clockwise* moment of magnitude

$$\delta M_r = \delta F_1 y_1 + \delta F_2 y_2$$

where $\quad \delta F_1 = \frac{\sigma_t}{y_t} a_1 y_1 \quad$ and $\quad \delta F_2 = \frac{\sigma_t}{y_t} a_2 y_2$

$$\therefore \quad \delta M_r = \frac{\sigma_t}{y_t} a_1 y_1 y_1 + \frac{\sigma_t}{y_t} a_2 y_2 y_2$$

$$= \frac{\sigma_t}{y_t} (a_1 y_1^2 + a_2 y_2^2)$$

$\therefore$ the total clockwise moment about the neutral axis is given by

$$M_r = \Sigma \delta M_r$$

$$= \frac{\sigma_t}{y_t} \Sigma a y^2$$

This moment is known as the *internal moment of resistance to bending* and, since the beam is in equilibrium, it is equal in magnitude but opposite in direction to the external bending moment M;

$$\therefore \quad M_r = M = \frac{\sigma_t}{y_t} \Sigma a y^2$$

The term $\Sigma a y^2$ is known as the *second moment of area*, I, about the neutral axis, and has units metres to the fourth (m^4). Thus, at *any* section y from the neutral axis,

$$M = \frac{\sigma}{y} I$$

or $\quad \sigma = \frac{M}{I} y$

From section 2.3,

$$\sigma = \frac{E}{R} y$$

$$\therefore \quad \frac{M}{I} = \frac{\sigma}{y} = \frac{E}{R}$$

which is known as the simple bending equation and should be remembered.

2.6 Second moment of area of common sections

For the *rectangle* (fig. 2.7(a)), the second moment of area about the neutral axis is given by

$$I_{\text{n.a.}} = \frac{bd^3}{12}$$

which should be remembered.

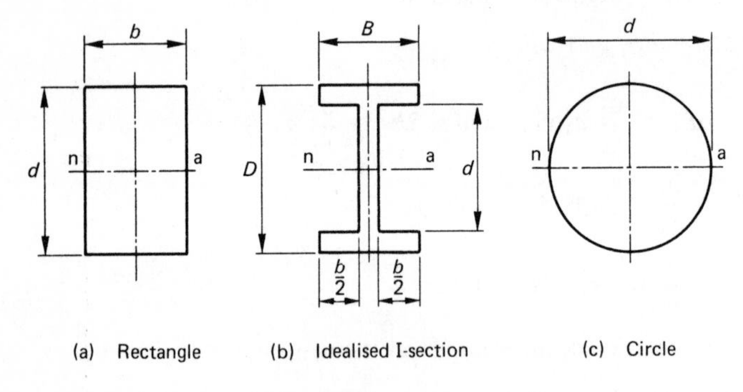

(a) Rectangle (b) Idealised I-section (c) Circle

Fig. 2.7 Second moment of area of common sections

For the *idealised I-section* (fig. 2.7(b)),

$$I_{\text{n.a.}} = \frac{BD^3}{12} - \frac{bd^3}{12}$$

$$= \frac{1}{12}(BD^3 - bd^3)$$

which it is useful to remember.

For the *circle* (fig. 2.7(c)),

$$I_{\text{n.a.}} = \frac{\pi d^4}{64}$$

which should be remembered.

2.7 Applications of the simple bending equation

Example 1 A beam of rectangular cross-section, 40 mm x 160 mm deep, is subjected to a bending moment of 35 kN m. Determine the maximum bending stress in the beam material.

From the simple bending equation,

$$\sigma = My/I$$

where $M = 35\,\text{kN m} = 35 \times 10^3\,\text{N m}$ $\quad y = 80\,\text{mm} = 0.08\,\text{m}$

and $I = bd^3/12 = 40\,\text{mm} \times (160\,\text{mm})^3/12 = 13.65 \times 10^6\,\text{mm}^4$

$= 13.65 \times 10^{-6}\,\text{m}^4$

(notice that, to convert mm^4 to m^4, multiply by 10^{-12})

$$\therefore \quad \sigma = \frac{35 \times 10^3\,\text{N m} \times 0.08\,\text{m}}{13.65 \times 10^{-6}\,\text{m}^4}$$

$$= 205.1 \times 10^6\,\text{N/m}^2 \quad \text{or} \quad 205.1\,\text{MN/m}^2$$

i.e. the bending stress is $205.1\,\text{MN/m}^2$.

Example 2 The cross-section of the idealised I-beam shown in fig. 2.8 has overall dimensions of 100 mm x 150 mm deep with a flange and web thickness of 20 mm. Calculate the second moment of area of the section about the neutral axis. If the bending stress in the beam material is not to exceed $180\,\text{MN/m}^2$, determine the maximum bending moment in the beam.

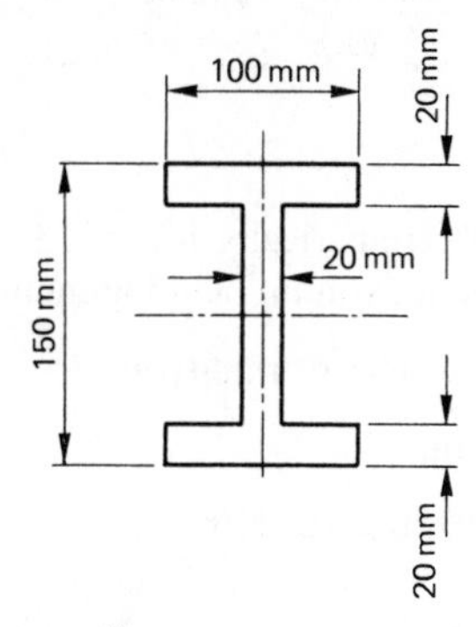

Fig. 2.8 Idealised I-beam section

For an idealised I-beam,

$$I_{\text{n.a.}} = \frac{1}{12}(BD^3 - bd^3)$$

where $B = 100\,\text{mm} = 0.1\,\text{m}$ $\quad D = 150\,\text{mm} = 0.15\,\text{m}$

$b = (100 - 20)\,\text{mm} = 0.08\,\text{m}$ and $d = (150 - 40)\,\text{mm} = 0.11\,\text{m}$

$$\therefore \quad I_{\text{n.a.}} = \frac{1}{12}[0.1\,\text{m} \times (0.15\,\text{m})^3 - 0.08\,\text{m} \times (0.11\,\text{m})^3]$$

$$= 1.925 \times 10^{-5}\,\text{m}^4$$

i.e. the second moment of area about the neutral axis is $1.925 \times 10^{-5}\,\text{m}^4$.

From the simple bending equation,

$$M = \sigma I / y$$

where $\sigma = 180\,\text{MN/m}^2 = 180 \times 10^6\,\text{N/m}^2$ $\quad y = 75\,\text{mm} = 0.075\,\text{m}$

and $I = 1.925 \times 10^{-5}\,\text{m}^4$

$$\therefore \quad M = \frac{180 \times 10^6\,\text{N/m}^2 \times 1.925 \times 10^{-5}\,\text{m}^4}{0.075\,\text{m}}$$

$$= 46.2 \times 10^3\,\text{Nm} \quad \text{or} \quad 46.2\,\text{kNm}$$

i.e. the maximum bending moment is 46.2 kNm.

Example 3 A steel bar, 100 mm diameter, is used to support a pulley block. If the bar is simply supported over a span of 1.2 m, determine the maximum bending stress in the material when the pulley block is at mid-span supporting a load of 20 kN.

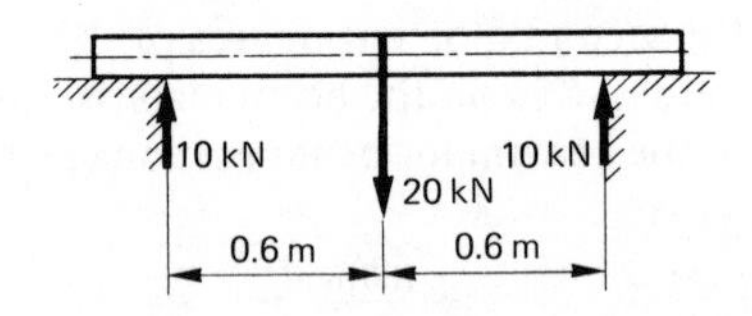

Fig. 2.9

The problem is shown diagrammatically in fig. 2.9.

Referring to fig. 2.9, the maximum bending moment occurs at mid-span,

$$\therefore \quad M_{max.} = \Sigma \text{ moments to } \textit{left} \text{ of mid-span}$$

$$= 10\,\text{kN} \times 0.6\,\text{m}$$

$$= 6\,\text{kN/m} \quad \text{or} \quad 6 \times 10^3\,\text{Nm}$$

$$\sigma_{max.} = M_{max.}\, y / I$$

where $y = (100\,\text{mm})/2 = 0.05\,\text{m}$

and $I = \pi d^4/64 = \pi(0.1\,\text{m})^4/64 = 4.91 \times 10^{-6}\,\text{m}^4$

$$\therefore \quad \sigma_{max.} = \frac{6 \times 10^3\,\text{Nm} \times 0.05\,\text{m}}{4.91 \times 10^{-6}\,\text{m}^4}$$

$$= 61.1 \times 10^6\,\text{N/m}^2 \quad \text{or} \quad 61.1\,\text{MN/m}^2$$

i.e. the maximum bending stress is 61.1 MN/m^2.

Example 4 The maximum bending moment and bending stress in a beam of rectangular cross-section must not exceed 25 kNm and 130 MN/m^2 respectively. What will be the minimum dimensions of the beam cross-section if the width of the beam is two-thirds of its depth?

$$I = bd^3/12 \quad \text{and} \quad b = \tfrac{2}{3}d$$

$$\therefore \quad I = \frac{2}{3}d \times \frac{d^3}{12} = \frac{d^4}{18}$$

From the simple bending equation,

$$\sigma = My/I$$

where $\sigma = 130\,\text{MN/m}^2 = 130 \times 10^6\,\text{N/m}^2$

$M = 25\,\text{kNm} = 25 \times 10^3\,\text{Nm}$ and $y = 0.5\,d$

$$\therefore \quad 130 \times 10^6\,\text{N/m}^2 = \frac{25 \times 10^3\,\text{Nm} \times 0.5\,d}{d^4/18}$$

$$\therefore \quad d^3 = \frac{25 \times 10^3\,\text{Nm} \times 0.5 \times 18}{130 \times 10^6\,\text{N/m}^2}$$

$$= 1.731 \times 10^{-3}\,\text{m}^3$$

$$\therefore \quad d = 0.12\,\text{m} \quad \text{or} \quad 120\,\text{mm}$$

and $b = \tfrac{2}{3} \times 120\,\text{mm}$

$= 80\,\text{mm}$

i.e. the beam dimensions are 80 mm x 120 mm deep.

Example 5 Calculate the maximum force which can be applied at mid-span to the simply supported beam shown in fig. 2.10, if the maximum bending stress at this point is not to exceed $200\,\text{MN/m}^2$.

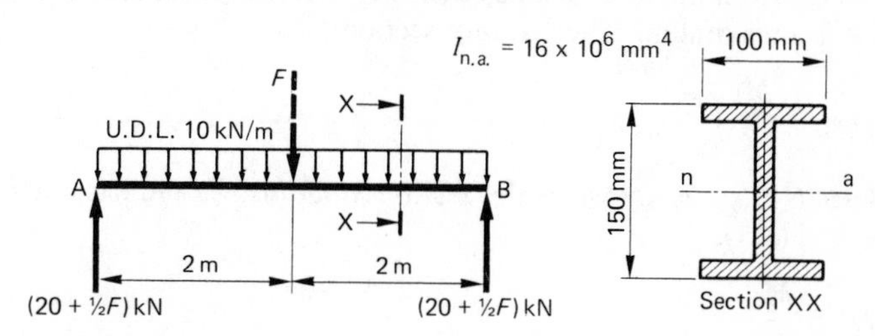

Fig. 2.10

By inspection of fig. 2.10,

reaction at A = reaction at B

$= \tfrac{1}{2}[(10\,\text{kN/m} \times 4\,\text{m}) + F]$

$= (20 + \tfrac{1}{2}F)\,\text{kN}$

where F is in kilonewtons.

At mid-span,

$$M = \Sigma \text{ moments to } \textit{left} \text{ of mid-span}$$
$$= [(20 + \tfrac{1}{2}F)\,\text{kN} \times 2\,\text{m}] - [(10\,\text{kN/m} \times 2\,\text{m}) \times 1\,\text{m}]$$
$$= (20 + F)\,\text{kNm} \quad \text{or} \quad (20 + F) \times 10^3\,\text{Nm}$$

From the simple bending equation,

$$M = \sigma I / y$$

where $\sigma = 200\,\text{MN/m}^2 = 200 \times 10^6\,\text{N/m}^2$

$I = 16 \times 10^6\,\text{mm}^4 = 16 \times 10^{-6}\,\text{m}^4$

and $y = (150\,\text{mm})/2 = 0.075\,\text{m}$

$$\therefore \quad (20 + F) \times 10^3\,\text{Nm} = \frac{200 \times 10^6\,\text{N/m}^2 \times 16 \times 10^{-6}\,\text{m}^4}{0.075\,\text{m}}$$
$$= 42.67 \times 10^3\,\text{N m}$$
$$\therefore \quad = 42.67 - 20$$
$$= 22.67\,\text{kN}$$

i.e. the maximum force at mid-span is 22.67 kN.

2.8 Section modulus

At any section in a loaded beam, the greatest bending stress occurs in the fibres furthest from the neutral axis of the beam cross-section, i.e. when the value of y is a maximum. Thus, at *any* section,

$$\sigma_{max.} = \frac{My_{max.}}{I}$$

The term $I/y_{max.}$ is known as the *section modulus*, Z, and has units metres cubed (m^3),

$$\therefore \quad \sigma_{max.} = \frac{M}{Z}$$

which it is useful to remember.

For beams of rectangular cross-section (b wide x d deep),

$$Z = \frac{I}{y_{max.}}$$
$$= \frac{bd^3/12}{d/2}$$
$$= \frac{bd^3}{6}$$

For circular beams of diameter d,

$$Z = \frac{\pi d^4/64}{d/2}$$

$$= \frac{\pi d^3}{32}$$

Example 1 Calculate the section modulus of a rectangular beam having a cross-section 200 mm wide by 300 mm deep.

$$Z = bd^3/6$$

where $b = 200\ \text{mm} = 0.2\ \text{m}$ and $d = 300\ \text{mm} = 0.3\ \text{m}$

$$\therefore \quad Z = \frac{0.2\ \text{m} \times (0.3\ \text{m})^3}{6}$$

$$= 0.0009\ \text{m}^3$$

i.e. the section modulus is $0.0009\ \text{m}^3$.

Example 2 What diameter of beam will have the same section modulus as the beam in example 1 above?

For a circular beam,

$$Z = \pi d^3/32$$

$$\therefore \quad d^3 = \frac{32Z}{\pi}$$

$$= \frac{32 \times 0.0009\ \text{m}^3}{\pi}$$

$$= 0.009\,17\ \text{m}^3$$

$$\therefore \quad d = 0.2093\ \text{m} \quad \text{or} \quad 209.3\ \text{mm}$$

i.e. the diameter is 209.3 mm.

Example 3 At a point in a beam having a section modulus of $21.3 \times 10^{-3}\ \text{m}^3$, the bending moment is 40 kNm. What is the maximum bending stress at this point?

$$\sigma_{\text{max.}} = M/Z$$

where $M = 40\ \text{kNm} = 40 \times 10^3\ \text{Nm}$ and $Z = 21.3 \times 10^3\ \text{m}^3$

$$\therefore \quad \sigma_{\text{max.}} = \frac{40 \times 10^3\ \text{Nm}}{21.3 \times 10^{-3}\ \text{m}^3}$$

$$= 1.88 \times 10^6\ \text{N/m}^2 \quad \text{or} \quad 1.88\ \text{MN/m}^2$$

i.e. the maximum bending stress is $1.88\ \text{MN/m}^2$.

2.9 Tables of properties

In general engineering practice, beams are selected from available standard stock sizes for which values of I, Z, etc. are known. A selection of readily available standard beam cross-sections are shown in Tables 2.1, 2.2, and 2.3. The values in these tables, which are abridged, have been reproduced from the *Structural steelwork handbook* by kind permission of the British Constructional Steelwork Association Limited and the Constructional Steel Research and Development Organisation.

Table 2.1 Universal beams (fig. 2.11)

Size $D \times B$ (mm)	Mass per metre (kg)	Thickness Web t (mm)	Thickness Flange T (mm)	I_{xx} (cm^4)	I_{yy} (cm^4)	Z_{xx} (cm^3)	Z_{yy} (cm^3)
910.3 x 304.1	222	15.9	23.9	375 924	11 223	8259	738.1
840.7 x 292.4	194	14.7	21.7	279 450	9069	6648	620.4
762.0 x 266.7	173	14.3	21.6	205 177	6846	5385	513.4
677.9 x 253.0	125	11.7	16.2	118 003	4379	3481	346.1
617.5 x 307.0	113	14.1	23.6	151 631	11 412	4911	743.3
533.1 x 209.3	92	10.2	15.6	55 353	2392	2076	228.6
467.4 x 192.8	98	11.4	19.6	45 717	2343	1956	243.0
457.2 x 151.9	67	9.1	15.0	28 577	878	1250	115.5
402.3 x 142.4	46	6.9	11.2	15 647	539	777.8	75.7
364.0 x 173.2	67	9.1	15.7	19 522	1362	1073	157.3
352.8 x 126.0	39	6.5	10.7	10 087	357	571.8	56.6
310.9 x 166.8	54	7.7	13.7	11 710	1061	753.3	127.3
312.7 x 102.4	33	6.6	10.8	6487	193	415.0	37.8
254.0 x 101.6	22	5.8	6.8	2867	120	225.7	23.6
203.2 x 133.4	25	5.8	7.8	2356	310	231.9	46.4

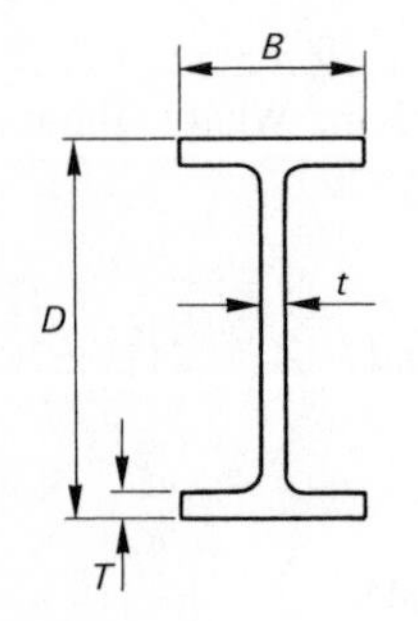

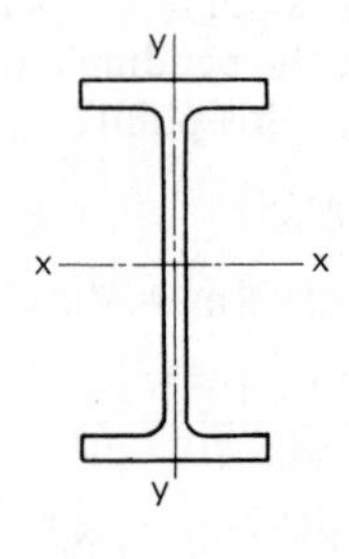

Fig. 2.11 Universal beam

Table 2.2 Structural tees cut from universal beams (fig. 2.12)

Size $B \times A$ (mm)	Mass per metre (kg)	Thickness: Web t (mm)	Thickness: Flange T (mm)	I_{xx} (cm^4)	Z_{xx} (cm^3)	Extreme fibre distance: y_f (cm)	Extreme fibre distance: y_t (cm)
304.1 x 455.2	112	15.9	23.9	29 036	869.8	12.1	33.42
292.4 x 420.4	97	14.7	21.7	21 381	690.9	11.1	30.94
266.7 x 381.0	87	14.3	21.6	15 495	551.3	9.99	28.11
253.0 x 339.0	63	11.7	16.2	8997	359.5	8.87	25.03
307.0 x 308.7	90	14.1	23.6	8949	369.5	6.65	24.22
209.3 x 266.6	46	10.2	15.6	3905	194.4	6.56	20.10
192.8 x 233.7	49	11.4	19.6	2978	167.2	5.55	17.82
151.9 x 228.6	34	9.1	15.0	2128	126.1	5.99	16.87
142.4 x 201.2	23	6.9	11.2	1130	75.0	5.05	15.07
173.2 x 182.0	34	9.1	15.7	1158	81.6	4.01	14.19
126.0 x 176.4	20	6.5	10.7	719.6	54.4	4.41	13.23
166.8 x 155.4	27	7.7	13.7	636.4	51.5	3.19	12.35
102.4 x 156.3	17	6.6	10.8	486.7	42.4	4.14	11.49
101.6 x 127.0	11	5.8	6.8	227.2	24.7	3.49	9.21
133.4 x 101.6	13	5.8	7.8	133.7	16.6	2.12	8.04

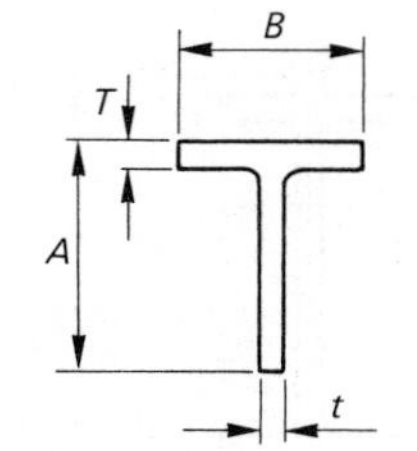

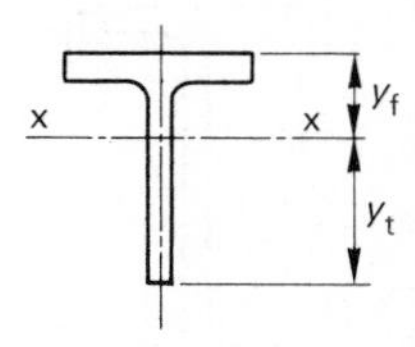

Fig. 2.12 Structural tee cut from universal beam

Example 1 The maximum tensile stress in a universal beam is limited to 80 N/mm^2 when the bending moment is 30 kN m. From Table 2.1, select a suitable size of beam, ignoring the effect of the mass of the beam itself and assuming the xx axis to be the neutral plane. What will be the actual stress in the chosen beam?

$$Z = M/\sigma_{max.}$$

where $M = 30\,\text{kNm} = 30 \times 10^3\,\text{Nm}$

and $\sigma_{max.} = 80\,\text{N/mm}^2 = 80 \times 10^6\,\text{N/m}^2$

Table 2.3 Gantry girders (fig. 2.13)

Size $D \times B$ (mm)	Mass per metre (kg)	Extreme fibre distance y_1 (cm)	y_2 (cm)	I_{xx} (cm^4)	Z_{xx} (cm^3)
939 x 432	355	39.20	54.68	644 504	11 797
853 x 432	260	32.92	52.37	384 673	7345
774 x 432	239	29.16	48.26	288 718	5982
700 x 432	218	25.58	44.40	215 582	4856
630 x 432	245	24.11	38.86	206 436	5313
627 x 381	195	23.60	39.14	155 882	3982
622 x 381	180	22.74	39.49	140 490	3558
618 x 305	155	23.53	38.22	120 871	3163
555 x 381	177	20.27	35.23	108 879	3091
550 x 305	151	20.84	34.13	92 644	2715
478 x 305	140	17.77	29.98	64 431	2149
472 x 254	117	18.67	28.50	54 593	1916
465 x 229	93	17.57	28.91	39 916	1381
423 x 305	116	14.67	27.62	40 572	1469
418 x 254	95	15.52	26.23	34 059	1298

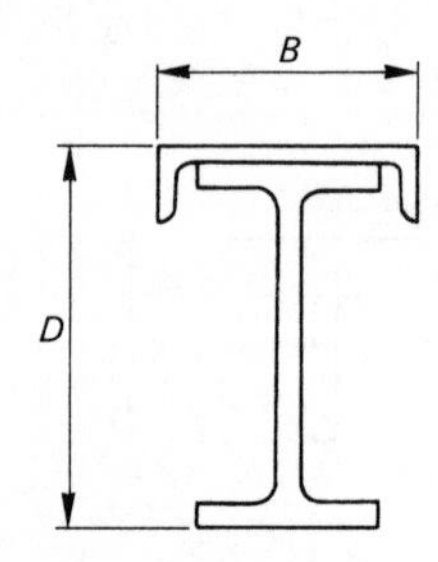

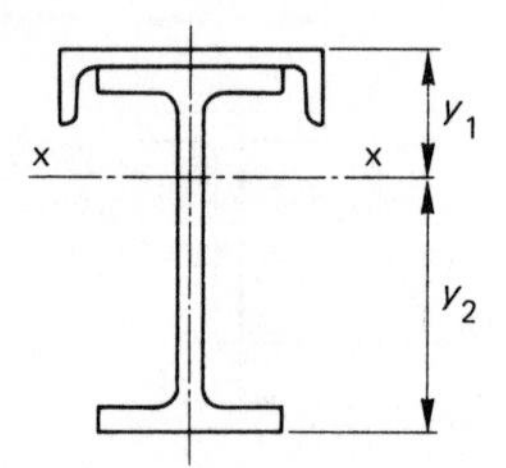

Fig. 2.13 Gantry girder

$$\therefore \quad Z = \frac{30 \times 10^3 \text{ Nm}}{80 \times 10^6 \text{ N/m}^2}$$

$$= 0.375 \times 10^{-3} \text{ m}^3$$

$$= 375 \text{ cm}^3$$

[note that values for Z in the tables are in units centimetres cubed (cm^3)]

It should be noted that,

$$1 \text{ m}^3 = 1 \times 10^{-6} \text{ cm}^3$$

In Table 2.1, $Z = 375\ \text{cm}^3$ lies between the Z_{XX} values 225.7 cm^3 and 415 cm^3. If the smaller value for Z_{XX} is used, the maximum stress will be exceeded,

i.e. $\sigma_{max.} = M/Z$

$$= \frac{30 \times 10^3\ \text{Nm}}{225.7 \times 10^{-6}\ \text{m}^3}$$

$$= 132.9 \times 10^6\ \text{N/m}^2 \quad \text{or} \quad 132.9\ \text{N/mm}^2$$

therefore, select the beam for which $Z_{XX} = 415\ \text{cm}^3$;

i.e. the selected beam is 312.7 mm deep x 102.4 mm wide.

Actual stress in chosen beam $= M/Z_{XX}$

$$= \frac{30 \times 10^3\ \text{Nm}}{415 \times 10^{-6}\ \text{m}^3}$$

$$= 72 \times 10^6\ \text{N/m}^2 \quad \text{or} \quad 72\ \text{N/mm}^2$$

i.e. the actual stress in the selected beam is 72 N/mm^2.

Example 2 A cantilever beam is 2 metres long and is to support a load of 10 kN at its free end. Select a suitable tee-section beam, indicating the correct orientation, to limit the maximum *tensile* stress in the beam material to 57 MN/m^2. Take into account the mass of the beam.

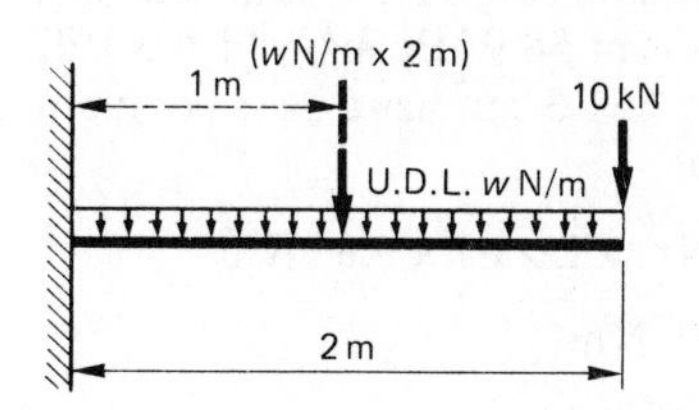

Fig. 2.14

The loaded beam is shown in fig. 2.14. Let the beam have weight w N/m,

i.e. w = mass per metre x g

Referring to fig. 2.14, the maximum bending moment will occur at the support,

$$\therefore \quad M_{max.} = -10\ \text{kN} \times 2\ \text{m} - (w\ \text{N/m} \times 2\ \text{m}) \times 1\ \text{m}$$

$$= -20 \times 10^3\ \text{Nm} - 2w\ \text{Nm}$$

The minus sign indicates that the maximum tensile bending stress is in the upper fibres of the beam cross-section and for the remaining calculations may be ignored; i.e. the beam should be loaded with the flange on the underside.

Since the mass of the beam is unknown, it is first necessary to calculate Z considering the bending moment due to the 10 kN force only (i.e. $M = 20 \times 10^3$ Nm).

$$Z = M/\sigma_{max.}$$

and $\sigma_{max.} = 57\ \text{MN/m}^2 = 57 \times 10^6\ \text{N/m}^2$

$$\therefore \quad Z = \frac{20 \times 10^3\ \text{Nm}}{57 \times 10^6\ \text{N/m}^2}$$

$$= 0.351 \times 10^{-3}\ \text{m}^3 \quad \text{or} \quad 351\ \text{cm}^3$$

From Table 2.2, the nearest larger Z value is 359.5 cm^3, which has a mass per metre of 63 kg;

$$\therefore \quad \text{maximum bending moment} = 20 \times 10^3\ \text{Nm} + 2 \times 63\ \text{kg} \times 9.81\ \text{m/s}^2 \times 1\ \text{m}$$

$$= 21.236 \times 10^3\ \text{Nm}$$

Check if the maximum bending stress has been exceeded:

$$\sigma_{max.} = M/Z$$

$$= \frac{21.236 \times 10^3\ \text{Nm}}{359.5 \times 10^{-6}\ \text{m}^3}$$

$$= 59.1 \times 10^6\ \text{N/m}^2$$

which is larger than the permissible stress.

Repeating the above procedure for a beam where $Z = 369.5$ cm^3 will result in a bending stress of 58.9 MN/m^2, which is still too large.

A beam where $Z = 551.3$ cm^3 has a mass per metre of 87 kg. Check for maximum stress value:

$$M = 20 \times 10^3\ \text{Nm} + 2 \times 87 \times 9.81\ \text{Nm}$$

$$= 21.71 \times 10^3\ \text{Nm}$$

and $$\sigma_{max.} = \frac{21.71 \times 10^3\ \text{Nm}}{551.3 \times 10^{-6}\ \text{m}^3}$$

$$= 39.4 \times 10^6\ \text{N/m}^2$$

which is within the permissible limit.
i.e. the beam selected is 266.7 mm wide x 381 mm deep and is used with the flange on the underside.

Example 3 A gantry girder, 618 mm deep x 305 mm wide, is used as a simply supported beam over a span of 8 metres and is to carry a uniformly distributed load. If the maximum tensile stress in the beam material is restricted to 163 N/mm^2, and taking account of the mass of the beam itself, determine the magnitude of the unknown uniformly distributed load. What will be the maximum compressive stress in the section?

From Table 2.3, for a gantry girder 618 mm x 305 mm,

$$\text{mass per metre} = 155\text{ kg} \qquad y_1 = 23.53\text{ cm} = 0.2353\text{ m}$$

$$I_{XX} = 120\,871\text{ cm}^4 = 120\,871 \times 10^{-8}\text{ m}^4$$

and $Z_{XX} = 3163\text{ cm}^3 = 3163 \times 10^{-6}\text{ m}^3$

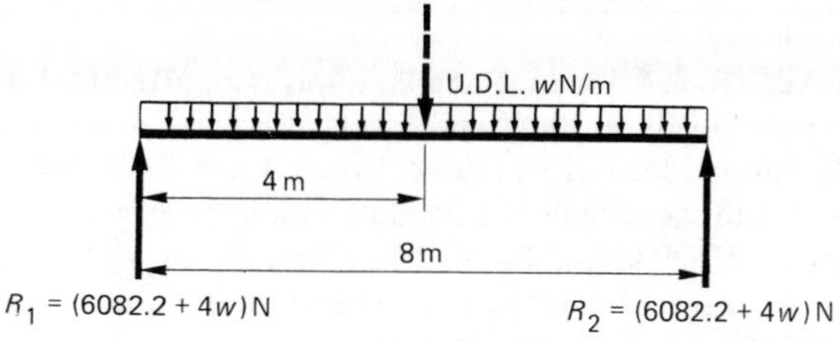

Fig. 2.15

Referring to fig. 2.15 and letting w N/m = unknown uniformly distributed load,

$$R_1 = R_2 = \frac{(155\text{ kg} \times 9.81\text{ m/s}^2 + w) \times 8\text{ m}}{2}$$

$$= (6082.2 + 4w)\text{ N}$$

The maximum bending moment will be at mid-span,

i.e. $M_{max.}$ = moments to *left* of mid-span

$$= (6082.2 + 4w)\text{ N} \times 4\text{ m} - (155 \times 9.81 + w)\text{ N/m} \times 4\text{ m} \times 2\text{ m}$$

$$= (12\,164.4 + 8w)\text{ Nm}$$

i.e. the beam is sagging and the maximum tensile stress will be in the flange on the underside of the beam.

Now $M_{max.} = Z\,\sigma_{max.}$

where $\sigma_{max.} = 163\text{ N/mm}^2 = 163 \times 10^6\text{ N/m}^2$

$$\therefore \quad 12\,164.4 + 8w = 3163 \times 10^{-6}\text{ m}^3 \times 163 \times 10^6\text{ N/m}^2$$

$$\therefore \quad 8w = 515\,569\text{ N m} - 12\,164\text{ Nm}$$

or $$w = 62.92 \times 10^3\text{ N/m}$$

i.e. the uniformly distributed load is 62.92 kN/m.

The maximum compressive stress will occur in the top flange,

$\therefore$ maximum compressive stress $\sigma_c = My_1/I_{xx}$

$$= \frac{(12\,164.4 + 8 \times 62.92 \times 10^3)\,\text{Nm} \times 0.2353\,\text{m}}{120\,871 \times 10^{-8}\,\text{m}^4}$$

$$= 100.4 \times 10^6\,\text{N/m}^2 \quad \text{or} \quad 100.4\,\text{N/mm}^2$$

i.e. the maximum compressive stress is 100.4 N/mm^2.

Exercises on chapter 2

1 A beam ABCDE is 4.5 m long and is simply supported at A and D. The beam carries point loads of 20 kN and 15 kN at B and E respectively, and a uniformly distributed load of 8 kN/m between A and C. If AB = 0.5 m and BC = CD = 1.5 m, determine the bending moments at A, B, C, D, and E. [$M_A = 0$; $M_B = 10.29$ kNm; $M_C = 2.15$ kNm; $M_D = -15$ kNm; $M_E = 0$]

2 A cantilever beam of rectangular cross-section is 2.2 m long and carries a point load of 8 kN at its free end. If the beam cross-section is 30 mm x 80 mm deep, determine the maximum bending stress in the material at a section 0.4 m from the support. [450 MN/m^2]

3 Steel strip, 4 mm thick, is to be wound on to a 1.6 m diameter drum. If the modulus of elasticity for the steel is 200 GN/m^2, determine the bending stress in the material. [250 MN/m^2]

4 A beam, 3 m long, is simply supported at its ends and is to carry a point load of 30 kN at mid-span. If the beam is rectangular in section, with its depth equal to twice its width, determine suitable dimensions to limit the maximum bending stress at mid-span to 80 MN/m^2. [75 mm x 150 mm]

5 The cross-section of an idealised I-section beam has overall dimensions 120 mm x 240 mm deep. If the web and flange are both 25 mm thick, determine the second moment of area of the section. If the maximum bending stress is limited to 100 MN/m^2, determine the maximum load the cross-section can support at mid-span when it is used as a beam 2.5 m long and simply supported at the ends. [83.94×10^{-6} m^4; 112 kN]

6 The flanges in an idealised I-section beam, 150 mm x 300 mm deep, are 25 mm thick and the web is 20 mm thick. Determine the dimensions of a rectangular-cross-section beam made from the same material, with a depth equal to twice the width, which will support the same maximum bending moment and bending stress. Compare the masses of the two beams. [238 mm x 119 mm; $\text{mass}_I = 0.441\ \text{mass}_{rectangle}$]

7 Working from first principles, show that the neutral axis of a beam cross-section which is symmetrical about its axis of loading passes through the centroid of the section.

A beam cross-section is tee-shaped, the axis of loading being along the axis of the vertical leg. If the horizontal 'cross bar' is 150 mm wide x 15 mm thick, and the vertical leg is 15 mm thick x 180 mm deep, calculate the position of the neutral axis relative to the top of the 'cross bar'. [60.68 mm]

8 A circular steel bar is used as a cantilever beam to carry a transverse force of 20 kN at a point 1.5 m from the support. Calculate (to the nearest millimetre) the diameter of the bar required to limit the maximum bending stress in the material to 170 MN/m^2. [122 mm]

9 Define the term 'section modulus'.

At a point in a beam with a section modulus of 0.4 x 10^{-3} m^3, the bending moment is 12.5 kN m. What is the bending stress? [31.25 MN/m^2]

10 The maximum bending stress and maximum bending moment in a beam cross-section are not to exceed 120 MN/m^2 and 30 kN m respectively. Calculate the section modulus. If $y_{max.}$ for the section is 160 mm, determine the second moment of area. [0.25 x 10^{-3} m^3; 40 x 10^{-6} m^4]

The following problems may be solved with the aid of the extracts from the tables of properties given in Tables 2.1, 2.2, and 2.3.

11 A universal beam, 254 mm deep x 101.6 mm wide, is to be used as a simply supported beam with a span of 4 m. If the bending stress at mid-span is restricted to 130 N/mm^2, what is the maximum load which can be applied to the beam at this point? [28.48 kN]

12 A gantry girder, 550 mm deep x 305 mm wide, is simply supported over a span of 18 m and carries a uniformly distributed mass of 300 kg per metre length together with point loads of 20 kN and 30 kN at 5 m and 12 m respectively from one support. Calculate the bending moment under each point load and at mid-span, taking into account the mass of the beam itself. Assuming that the maximum bending moment occurs at one of these points, calculate the maximum tensile and compressive bending stresses in the beam material. [M_{20} = 271.5 kN m; M_{30} = 325.9 kN m; M_{ms} = 329 kN m; 121.2 MN/m^2 (tensile); 74 MN/m^2 (compressive)]

13 A universal beam, ABCD, is simply supported at B and D and carries point loads of 45 kN at A and 60 kN at C. If AB = 1 m, BC = 3 m, and CD = 2 m, calculate the bending moments at A, B, C, and D. Assuming that the maximum bending moment occurs at one of these points, select from Table 2.1 the minimum size of beam which will restrict the maximum stress in the beam material to 120 N/mm^2. Ignore the mass of the beam itself. [352.8 mm deep x 126 mm wide]

14 A structural tee-section beam is to be used as a cantilever 3 m long *with the flange uppermost.* If the maximum tensile bending stress in the beam material at the support is restricted to 80 N/mm^2 when the load acting downwards at the free end is 4.5 kN, select the minimum size of beam. Take into account the mass of the beam itself. What will be the maximum compressive stress in the beam material? [166.8 mm wide x 155.4 mm deep; 285 MN/m^2]

3 Torsion of circular shafts

3.1 Shear stress and angle of twist

If a tangential force F is applied at D to the circular shaft of radius r shown in fig. 3.1(a), then a torque of magnitude $T = Fr$ will be applied about B. At the end A, which is rigidly held, there will be an equal and opposite torque acting which can be represented by the tangential force at C as shown in fig. 3.1(b). Since the lines of action of the tangential forces at C and D are separated, these forces are *shearing forces* (see section 1.1) and the angular displacement γ (*gamma*) of the line CD to CE is the *shear strain* at radius r over length l.

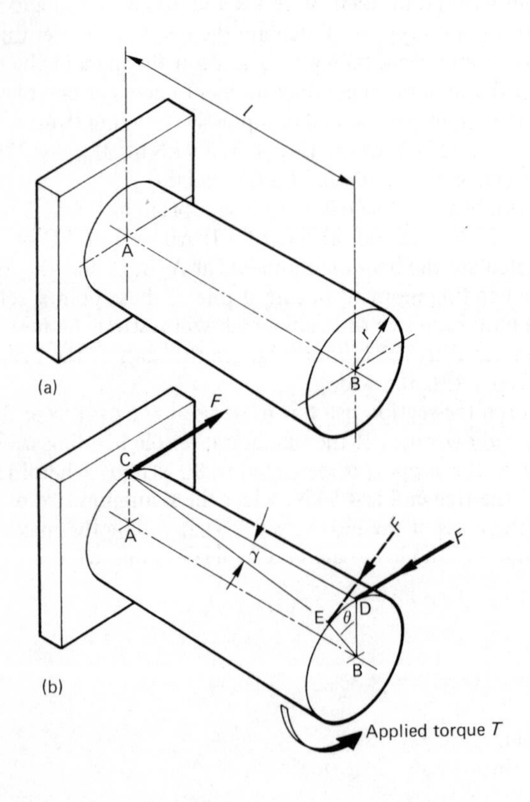

Fig. 3.1 Torsion of a circular shaft

Referring to fig. 3.1(b),

$$\text{shear strain } \gamma = \frac{DE}{l} \quad \text{(provided } \gamma \text{ is small)}$$

but $DE = r\theta$

(where θ is measured in radians)

$$\therefore \quad \gamma = \frac{r\theta}{l} \qquad \text{(i)}$$

i.e. shear strain = radius x twist per unit length

From section 1.10,

$$\text{shear modulus} = \frac{\text{shear stress}}{\text{shear strain}}$$

i.e.

$$G = \frac{\tau}{\gamma}$$

$$\therefore \quad \gamma = \frac{\tau}{G} \qquad \text{(ii)}$$

Equating (i) and (ii),

$$\frac{\tau}{G} = \frac{r\theta}{l}$$

or

$$\frac{\tau}{r} = \frac{G\theta}{l}$$

which should be remembered.

If it is assumed that the shaft remains perfectly circular during twisting and there is no change in the length or diameter, then, after twisting, the radial line BE in fig. 3.1(b) will be perfectly straight, and the *angle of twist* θ between the centre of the shaft, B, and the point on the circumference, E, will be constant – which is a reasonable assumption if the twist per unit length is small (typically 1° in a length equal to 15 diameters) – i.e., for small values of twist per unit length, the term $G\theta/l$ will be constant.

$$\therefore \quad \frac{\tau}{r} = \text{constant}$$

If τ_1, τ_2, and τ_n are the shear stresses at radii r_1, r_2, and r_n etc., then

$$\frac{\tau}{r} = \frac{\tau_1}{r_1} = \frac{\tau_2}{r_2} = \frac{\tau_n}{r_n} = \frac{G\theta}{l}$$

which it is useful to remember.

Example When subjected to an applied torque, a circular shaft of length 600 mm twists through an angle of 2°. If the shear modulus for the shaft material is 80 GN/m², determine the shear stress (a) at the centre of the shaft, (b) at a radius of 20 mm.

If the maximum shear stress in the shaft is 120 MN/m², what will be the outside diameter?

$$\frac{\tau}{r} = \frac{G\theta}{l}$$

where $G = 80\ \text{GN/m}^2 = 80 \times 10^9\ \text{N/m}^2$ $\theta = 2^\circ = 0.035\ \text{rad}$

and $l = 600\ \text{mm} = 0.6\ \text{m}$

(notice that the angle of twist, θ, must be converted to radians)

$$\therefore \quad \frac{\tau}{r} = \frac{80 \times 10^9\ \text{N/m}^2 \times 0.035\ \text{rad}}{0.6\ \text{m}}$$

$$= 4.67 \times 10^9\ \text{N/m}^3$$

a) At the centre of the shaft, $r = 0$
i.e. the shear stress at the centre of the shaft is zero.

b) When $r = 20\ \text{mm} = 0.02\ \text{m}$,

$$\tau = 4.67 \times 10^9\ \text{N/m}^3 \times 0.02\ \text{m}$$

$$= 93.4 \times 10^6\ \text{N/m}^2 \quad \text{or} \quad 93.4\ \text{MN/m}^2$$

i.e. the shear stress at 20 mm radius is 93.4 MN/m².

For a shear stress of 120 MN/m²,

$$\tau/r = 4.67 \times 10^9\ \text{N/m}^3$$

$$\therefore \quad r = \frac{\tau}{4.67 \times 10^9\ \text{N/m}^3}$$

where $\tau = 120\ \text{MN/m}^2 = 120 \times 10^6\ \text{N/m}^2$

$$\therefore \quad r = \frac{120 \times 10^6\ \text{N/m}^2}{4.67 \times 10^9\ \text{N/m}^3}$$

$$= 25.7 \times 10^{-3}\ \text{m} \quad \text{or} \quad 25.7\ \text{mm}$$

$$\therefore \quad \text{diameter} = 2\,r$$

$$= 2 \times 25.7\ \text{mm}$$

$$= 51.4\ \text{mm}$$

i.e., since the maximum shear stress occurs at maximum radius, the outside diameter of the shaft is 51.4 mm.

3.2 Relationship between shear stress and external torque

Figure 3.2(a) shows the cross-section of a circular shaft subjected to an external torque T, and fig. 3.2(b) illustrates the variation of shear stress across the section. The shear stress varies from zero at the centre to a maximum value at the outside diameter.

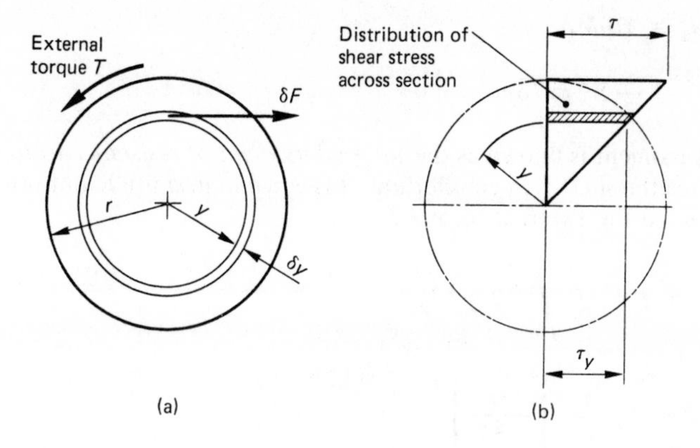

Fig. 3.2 Distribution of shear stress

Referring to fig. 3.2(a), the tangential force δF at radius y due to shear stress τ_y is given by

$$\delta F = \tau_y \times \text{area resisting shear}$$

and area resisting shear = area of elemental ring

$$= 2\,\pi y \delta y$$

$$\therefore \quad \delta F = \tau_y \times 2\,\pi y \delta y$$

From fig. 3.2(b),

$$\frac{\tau_y}{y} = \frac{\tau}{r}$$

$$\therefore \quad \tau_y = \frac{\tau y}{r}$$

and
$$\delta F = \frac{\tau y}{r} \times 2\,\pi y \delta y$$

$$= \frac{\tau}{r} \times 2\,\pi y^2 \delta y$$

The clockwise moment of this force about the centre of the shaft is given by

$$\delta M_{\mathrm{r}} = \frac{\tau}{r} \times 2\,\pi y^2 \delta y \times y$$

$$= \frac{\tau}{r} \times 2\,\pi y^3 \delta y$$

$\therefore$ the total clockwise moment about the centre of the shaft is given by

$$M_r = \Sigma \delta M_r$$

$$= \frac{\tau}{r} \Sigma 2\pi y^3 \delta y$$

This moment is known as the *internal moment of resistance to torsion*, and, since the shaft is in equilibrium, it is equal in magnitude but opposite in direction to the external torque T.

$$\therefore \quad M_r = T = \frac{\tau}{r} \int_0^r 2\pi y^3 \, dy$$

$$= \frac{\tau}{r} \left[\frac{2\pi y^4}{4} \right]_0^r$$

$$\therefore \quad T = \frac{\tau}{r} \times \frac{\pi r^4}{2}$$

The term $\pi r^4/2$ is known as the *polar second moment of area, J,* about the shaft axis and has units of metres to the fourth (m^4);

i.e. $J = \dfrac{\pi r^4}{2} = \dfrac{\pi d^4}{32}$ (since $d = 2r$)

which should be remembered.

$$\therefore \quad T = \frac{\tau}{r} J$$

or $\dfrac{T}{J} = \dfrac{\tau}{r}$

From section 3.1,

$$\frac{\tau}{r} = \frac{G\theta}{l}$$

$$\therefore \quad \frac{T}{J} = \frac{G\theta}{l} = \frac{\tau}{r}$$

which is known as the simple torsion equation and should be remembered.

Example 1 The feed drive shaft on a lathe is 20 mm diameter and is required to transmit a maximum torque of 1 N m to the apron. Neglecting the effect of the keyway or spline, determine the angle of twist and the maximum shear stress in the drive shaft when the effective driving length is 1.8 m. Take $G = 79\ GN/m^2$.

From the simple torsion equation,

i.e. $T/J = G\theta/l = \tau/r$

$\theta = Tl/GJ$ and $\tau = Tr/J$

where $T = 1\,\text{Nm}$ $\quad l = 1.8\,\text{m}$ $\quad G = 79\,\text{GN/m}^2 = 79 \times 10^9\,\text{N/m}^2$

$r = 10\,\text{mm} = 0.01\,\text{m}$

and $J = \pi d^4/32 = \pi \times (0.02\,\text{m})^4/32 = 1.57 \times 10^{-8}\,\text{m}^4$.

$$\therefore \quad \text{angle of twist } \theta = \frac{1\,\text{Nm} \times 1.8\,\text{m}}{79 \times 10^9\,\text{N/m}^2 \times 1.57 \times 10^{-8}\,\text{m}^4}$$

$$= 1.45 \times 10^{-3}\,\text{rad}$$

$$= 0.083^\circ \quad \text{or} \quad 5'$$

i.e. the angle of twist is 5 minutes.

The maximum shear stress occurs at the maximum radius,

$$\therefore \quad \tau_{\text{max.}} = \frac{1\,\text{Nm} \times 0.01\,\text{m}}{1.57 \times 10^{-8}\,\text{m}^4}$$

$$= 637 \times 10^3\,\text{N/m}^2 \quad \text{or} \quad 637\,\text{kN/m}^2$$

i.e. the maximum shear stress is $637\,\text{kN/m}^2$.

Example 2 In a shearing test on a material, failure occurred when the shear stress was $150\,\text{N/mm}^2$. Using a factor of safety of 4, determine the minimum diameter of a shaft made from the material which will transmit a power of 10 kW when rotating at 1500 rev/min. If the shaft is 1.2 m long, what will be the angle of twist? Take $G = 79\,\text{GN/m}^2$.

It is first necessary to determine the torque from the relationship

power = torque x angular velocity

i.e. $P = T\omega$

$\therefore$ $T = P/\omega$

where $P = 10\,\text{kW} = 10 \times 10^3\,\text{W}$

$$\text{and} \quad \omega = \frac{1500\,\text{rev/min} \times 2\pi\,\text{rad/rev}}{60\,\text{s/min}} = 157.1\,\text{rad/s}$$

$$\therefore \quad T = \frac{10 \times 10^3\,\text{W}}{157.1\,\text{rad/s}} = 63.65\,\text{Nm}$$

From the simple torsion equation,

i.e. $T/J = G\theta/l = \tau/r$

$T/\tau = J/r$

$$\text{where} \quad \tau = \frac{\text{shear stress at failure}}{\text{factor of safety}} = \frac{150\,\text{N/mm}^2}{4} = 37.5 \times 10^6\,\text{N/m}^2$$

$J = \pi d^4/32$ and $r = d/2$

$$\therefore \quad \frac{63.65\ \text{Nm}}{37.5 \times 10^6\ \text{N/m}^2} = \frac{\pi d^4/32}{d/2} = \frac{\pi d^3}{16}$$

$$\therefore \quad d^3 = \frac{63.65\ \text{Nm} \times 16}{37.5 \times 10^6\ \text{N/m}^2 \times \pi}$$

$$= 8.64 \times 10^{-6}\ \text{m}^3$$

$$\therefore \quad d = 0.0205\ \text{m} \quad \text{or} \quad 20.5\ \text{mm}$$

i.e. the minimum shaft diameter is 20.5 mm.

For this shaft diameter and maximum shear stress,

$$\text{angle of twist } \theta = \tau l / rG$$

where $l = 1.2\ \text{m}$ $\quad G = 79\ \text{GN/m}^2 = 79 \times 10^9\ \text{N/m}^2$

and $\quad r = (0.0205\ \text{m})/2 = 0.010\,25\ \text{m}$

$$\therefore \quad \theta = \frac{37.5 \times 10^6\ \text{N/m}^2 \times 1.2\ \text{m}}{0.010\,25\ \text{m} \times 79 \times 10^9\ \text{N/m}^2}$$

$$= 0.056\ \text{rad}$$

$$= 3.21^\circ$$

i.e. the angle of twist in a length of 1.2 m is 3.21°.

3.3 Torsion of hollow circular shafts

Figure 3.3(a) shows the cross-section of a hollow circular shaft subjected to an external torque T, and fig. 3.3(b) illustrates the distribution of shear stress across the section. The shear stress varies from a minimum at the bore to a maximum value at the outside diameter.

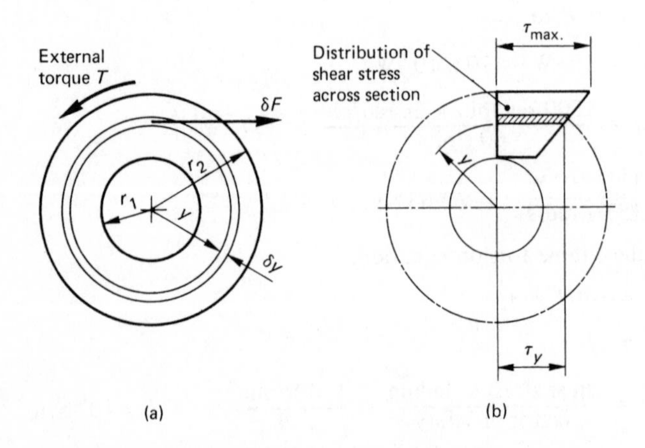

Fig. 3.3 Distribution of shear stress in a hollow shaft

From section 3.2, the total internal moment of resistance to torsion across the section is given by

$$M_r = T = \frac{\tau}{r} \Sigma\, 2\pi y^3 \delta y$$

and, for the section shown in fig. 3.3(a),

$$T = \frac{\tau}{r} \int_{r_1}^{r_2} 2\pi y^3 \, dy$$

$$= \frac{\tau}{r} \times \frac{\pi}{2} (r_2^4 - r_1^4)$$

The term $(\pi/2)\,(r_2^4 - r_1^4)$ is the polar second moment of area, J, for a hollow section; i.e., for a hollow shaft,

$$J = \frac{\pi}{2} (r_2^4 - r_1^4) = \frac{\pi}{32} (d_2^4 - d_1^4)$$

which should be remembered.

$$\therefore \quad T = \frac{\tau}{r} J$$

or $$\frac{T}{J} = \frac{\tau}{r}$$

From section 3.1,

$$\frac{\tau}{r} = \frac{G\theta}{l}$$

$$\therefore \quad \frac{T}{J} = \frac{G\theta}{l} = \frac{\tau}{r}$$

which is the simple torsion equation.

Example 1 A hollow shaft of external diameter 50 mm and bore 35 mm is to transmit a torque of 70 N m. If the shaft is 0.8 m long and the shear modulus is 80 GN/m², determine (a) the maximum shear stress, (b) the maximum angle of twist.

From the simple torsion equation,

i.e. $T/J = G\theta/l = \tau/r$

$\tau = Tr/J$ and $\theta = Tl/JG$

where $T = 70\,\text{N m}$ $\quad l = 0.8\,\text{m}$ $\quad G = 80\,\text{GN/m}^2 = 80 \times 10^9\,\text{N/m}^2$

and $J = \frac{\pi}{32} (d_2^4 - d_1^4) = \frac{\pi}{32} [(0.05\,\text{m})^4 - (0.035\,\text{m})^4] = 4.66 \times 10^{-7}\,\text{m}^4$

a) The maximum shear stress occurs at the outside diameter, i.e. when r = 0.025 m,

$$\therefore \quad \tau_{max.} = \frac{70\,\text{Nm} \times 0.025\,\text{m}}{4.66 \times 10^{-7}\,\text{m}^4}$$

$$= 3.76 \times 10^6\,\text{N/m}^2 \quad \text{or} \quad 3.76\,\text{MN/m}^2$$

i.e. the maximum shear stress is 3.76 MN/m².

b) Angle of twist $\theta = Tl/JG$

$$= \frac{70\,\text{Nm} \times 0.8\,\text{m}}{4.66 \times 10^{-7}\,\text{m}^4 \times 80 \times 10^9\,\text{N/m}^2}$$

$$= 0.0015\,\text{rad}$$

$$= 0.086^\circ \quad \text{or} \quad 5.2'$$

i.e. the maximum angle of twist is 5.2 minutes.

Example 2 The hollow shaft in example 1 is to be replaced by a solid circular shaft of equal length and made from the same material. If for an applied torque of 70 Nm the maximum shear stress and angle of twist are not to exceed the values found for the hollow shaft, calculate the minimum solid-shaft diameter. Compare the masses of the two shafts.

From the simple torsion equation,

i.e. $T/J = G\theta/l = \tau/r$

$T/\tau = J/r$ and $J = Tl/G\theta$

Thus, if the solid and hollow shafts are made from the same material and are subjected to the same torque which produces identical values for shear stress and angle of twist, then

$$\frac{J_s}{r_s} \text{ (solid shaft)} = \frac{J_h}{r_h} \text{ (hollow shaft)} \qquad \text{(i)}$$

and J_s (solid shaft) = J_h (hollow shaft) (ii)

Consider equation (i):

for the hollow shaft, $J_h = 4.66 \times 10^{-7}\,\text{m}^4$ and $r_h = 0.025\,\text{m}$

$$\therefore \quad \frac{J_h}{r_h} = \frac{4.66 \times 10^{-7}\,\text{m}^4}{0.025\,\text{m}} = 1.864 \times 10^{-5}\,\text{m}^3$$

for the solid shaft, $J_s = \pi d_s^4/32$ and $r_s = d_s/2$

$$\therefore \quad \frac{J_s}{r_s} = \frac{\pi d_s^4/32}{d_s/2} = \frac{\pi d_s^3}{16}$$

$$\therefore \quad \pi d_s^3/16 = 1.864 \times 10^{-5}\,\text{m}^3$$

$\therefore \quad d_s^3 = 9.5 \times 10^{-5}\ \mathrm{m}^3$

$\therefore \quad d_s = 0.0456\ \mathrm{m} \quad \text{or} \quad 45.6\ \mathrm{mm}$

Consider equation (ii):

for the solid shaft, $J_s = \pi d_s^4/32$

$\therefore \quad \pi d_s^4/32 = 4.66 \times 10^{-7}\ \mathrm{m}^4$

$\therefore \quad d_s^4 = 4.75 \times 10^{-6}\ \mathrm{m}^4$

$\therefore \quad d_s = 0.0467\ \mathrm{m} \quad \text{or} \quad 46.7\ \mathrm{mm}$

i.e. the minimum diameter of shaft which will satisfy *all* conditions is 46.7 mm.

Let m_s and m_h be the masses of the solid and hollow shafts respectively,

then $\dfrac{m_s}{m_h} = \dfrac{\text{density x volume of solid shaft}}{\text{density x volume of hollow shaft}}$

Since the shafts are made from the same material and are of equal length,

$$\frac{m_s}{m_h} = \frac{(\pi/4)\,(0.0467\ \mathrm{m})^2}{(\pi/4)\,(0.05^2 - 0.035^2)\ \mathrm{m}^2}$$

$$= 1.71$$

or $\quad m_s = 1.71\, m_h$

i.e. the mass of the solid shaft is 1.71 times greater than the mass of the hollow shaft. Thus, where the *mass* of the shaft is important, hollow shafts are used. For this reason, the propshaft in a front-engined, rear-wheel-drive motor vehicle is hollow.

Example 3 A hollow steel shaft is required to transmit 3000 kW when rotating at 120 rev/min. If the ratio of internal to external diameter is 1:3 and the shear stress is limited to 60 MN/m², determine the required diameters and the twist per unit length. For steel, $G = 79\ \mathrm{GN/m^2}$.

Power $P = T\omega$

$\therefore \quad T = P/\omega$

where $P = 3000\ \mathrm{kW}$ and $\omega = \dfrac{120\ \text{rev/min} \times 2\pi\ \text{rad/rev}}{60\ \text{s/min}} = 12.57\ \text{rad/s}$

$$\therefore \quad T = \frac{3000\ \mathrm{kW}}{12.57\ \mathrm{rad/s}}$$

$$= 238.7\ \mathrm{kNm}$$

$$= 238.7 \times 10^3\ \mathrm{Nm}$$

From the simple torsion equation,

i.e. $T/J = G\theta/l = \tau/r$

$J/r = T/\tau$

Let d_1 = internal diameter and d_2 = external diameter

then $d_2 = 3d_1$

thus $J = (\pi/32) \times [d_2^4 - d_1^4] = (\pi/32) \times [(3d_1)^4 - d_1^4] = 7.85d_1^4$

$r = d_2/2 = 3d_1/2$ and $\tau = 60\ \text{MN/m}^2 = 60 \times 10^6\ \text{N/m}^2$

$$\therefore \quad \frac{7.85d_1^4}{(3d_1/2)} = \frac{238.7 \times 10^3\ \text{Nm}}{60 \times 10^6\ \text{N/m}^2} = 3.98 \times 10^{-3}\ \text{m}^3$$

$$\therefore \quad d_1^3 = \frac{3.98 \times 10^{-3}\ \text{m}^3 \times 3}{7.85 \times 2}$$

$$= 7.6 \times 10^{-4}\ \text{m}^3$$

$$\therefore \quad d_1 = 0.0913\ \text{m} \quad \text{or} \quad 91.3\ \text{mm}$$

and $d_2 = 91.3\ \text{mm} \times 3$

$= 273.9\ \text{mm}$

i.e. the internal and external diameters are 91.3 mm and 273.9 mm respectively.

Now, $\theta/l = \tau/Gr$

where $G = 79\ \text{GN/m}^2 = 79 \times 10^9\ \text{N/m}^2$

and $r = (0.2739\ \text{m})/2 = 0.137\ \text{m}$

$$\therefore \quad \frac{\theta}{l} = \frac{60 \times 10^6\ \text{N/m}^2}{79 \times 10^9\ \text{N/m}^2 \times 0.137\ \text{m}}$$

$$= 5.54 \times 10^{-3}\ \text{rad/m}$$

$$= 0.317°/\text{m} \quad \text{or} \quad 19\ \text{minutes/m}$$

i.e. the angle of twist per unit length is 19 minutes/m.

Exercises on chapter 3

1 A solid circular shaft is 50 mm diameter and 2 m long. Determine the maximum shear stress and angle of twist when the shaft is subjected to an applied torque of 85 Nm. Take G as 80 GN/m^2. [3.46 MN/m^2; 11.9′]

2 A solid circular shaft is required to transmit 15 kW when rotating at 3000 rev/min. If the twist per unit length and the maximum shear stress are not to exceed 1.5 degrees/metre and 50 MN/m^2 respectively, determine the minimum diameter of shaft required. G = 80 GN/m^2. [21.9 mm]

3 Working from first principles, develop the simple torsion equation.

A shaft, 100 mm diameter and 470 mm long, transmits 4 kW when rotating at 1500 rev/min. If the shear modulus is 60 kN/mm², determine the maximum shear stress and angle of twist in the shaft material.
[20.35 N/mm²; 11′]

4 The torque on a rotating shaft was determined by measuring the angle of twist. If the diameter of the shaft under test was 70 mm and the twist per unit length was 0.5 degrees/metre, determine the applied torque. What power is the shaft transmitting if it is rotating at 600 rev/min? Take $G = 78$ GN/m².
[1.6 kN m; 100.5 kW]

5 The maximum shear stress and angle of twist in a shaft material are not to exceed 40 N/mm² and 2° respectively. If the diameter of the shaft is 45 mm and the shear modulus is 79 kN/mm², determine the maximum length of the shaft. How much power will the shaft transmit when it is rotating at 900 rev/min? [1.555 m; 67.45 kW]

6 A hollow shaft of internal and external diameters 50 mm and 80 mm respectively is 2 m long. If the maximum shear stress in the shaft material is not to exceed 55 MN/m², determine the power the shaft will transmit when it is rotating at 2000 rev/min. What will be the angle of twist?
Take $G = 80$ GN/m². [11.8 MW; 1.97°]

7 A hollow shaft having an external to internal diameter ratio of 2:1 is subjected to a torque of 1500 N m. If the maximum shear stress in the shaft material is not to exceed 70 MN/m², determine the shaft dimensions.
[O/D 48.8 mm; bore 24.4 mm]

8 If the shaft in question 7 is replaced by a solid shaft having the same external diameter and equal length, determine (a) the maximum shear stress when the solid shaft is subjected to a torque of 1500 N m, (b) the percentage increase in the mass of the shaft. [65.7 MN/m²; 33%]

9 The maximum shear stress in a hollow shaft is not to exceed 60 MN/m². If the ratio of internal to external diameter is 1:2, show that the external diameter is given by

$$D = \sqrt[3]{\frac{P}{1156N}} \text{ metres}$$

where P = power measured in kilowatts and N = shaft speed in rev/min.

Determine the dimensions of such a shaft to transmit 9000 kW at 600 rev/min, and find the angle of twist per unit length.
Take $G = 80$ GN/m². [235 mm; 117.5 mm; 22 min/m]

10 Why is it advantageous to use a hollow shaft instead of a solid shaft to transmit power?

A solid shaft, 80 mm diameter, is to be replaced by a hollow shaft made from the same material and of equal length. If the external diameter of the hollow shaft is 100 mm and the maximum shear stress in both shafts is to remain the same for the same applied torque, calculate the diameter of the bore. For the two shafts, compare (a) the masses, (b) the angles of twist.
[83.6 mm; $m_s/m_h = 2.13$; $\theta_s/\theta_h = 1.25$]

11 A solid circular shaft of diameter 20 mm and length 375 mm is used as a torsion indicator by measuring the angle of twist. If the maximum angle of twist for an applied torque T is 1.5°, find the magnitude of T. To increase the sensitivity of the device so that for the same applied torque T the angle of twist is 3°, the shaft is replaced by a hollow shaft made from the same material and having the same external dimensions. What will be the bore size? G = 60 GN/m². [65.8 Nm; 16.8 mm]

12 A hollow shaft with external diameter 100 mm and bore 60 mm is 1.6 m long. If the maximum shear stress and angle of twist are not to exceed 60 N/mm² and 2° respectively, determine the maximum speed of rotation of the shaft when transmitting 900 kW. Take G = 80 kN/mm². [838 rev/min]

13 The following results were obtained from a torsion test on a metallic sample of diameter 6 mm using a torsion meter with a gauge length of 50 mm:

Applied torque (Nm)	2	4	6	8	10	12	14	16	18
Angle of twist (rad)	0.013	0.025	0.039	0.053	0.066	0.078	0.092	0.112	0.130

Plot a graph of torque against angle of twist and use it to determine the shear modulus for the material. [60 GN/m²]

14 The angle of twist in a shaft is restricted to one degree in a length equal to twenty diameters. If the shear modulus in such a shaft is 80 GN/m², determine the shear stress. [34.9 MN/m²]

15 A hollow shaft with a ratio of outside diameter to bore of 4:3 is to transmit a mean power of 100 kW while rotating at 400 rev/min. If the maximum torque is 1.36 times the mean torque and the shear stress is not to exceed 80 N/mm², determine the dimensions of the shaft.
[67.12 mm x 50.34 mm]

16 Determine the maximum shear stress in a steel propeller shaft, 400 mm external diameter and 200 mm internal diameter, when subjected to a torque of 450 kNm. If the shear modulus of steel is 80 kN/mm², find the angle of twist in a length equal to 20 times the external diameter. Determine the diameter of a replacement solid shaft manufactured from the same material and subjected to the same maximum shear stress.
[38.2 N/mm²; 1.09°; 181.7 mm]

4 Angular motion

4.1 Equations of uniform angular motion

Referring to the graph of angular velocity against time shown in fig. 4.1,

ω_1 = initial angular velocity, with units radians per second (rad/s);

ω_2 = final angular velocity, with units rad/s;

α = uniform or constant angular acceleration, with units rad/s^2

and t = time, with units seconds (s).

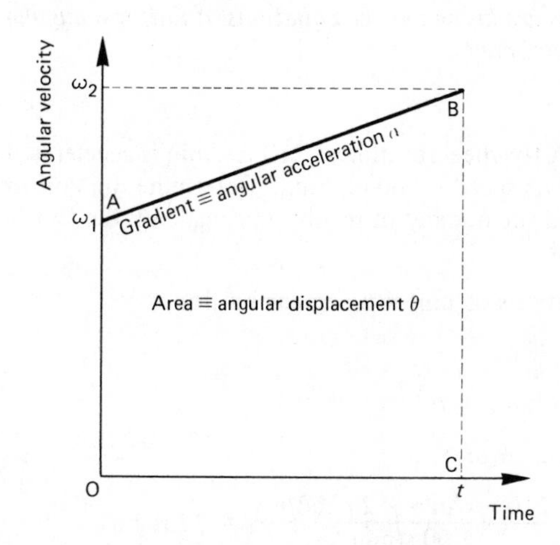

Fig. 4.1 Graph of angular velocity against time

Angular displacement $\theta \equiv$ area OABC

i.e. $\theta = \frac{1}{2}(\omega_2 + \omega_1)t$ (i)

and angular acceleration $\alpha \equiv$ gradient or slope of line AB

i.e. $\alpha = \dfrac{\omega_2 - \omega_1}{t}$ (ii)

Multiplying equation (i) by equation (ii) gives

$$\alpha\theta = \frac{1}{2}(\omega_2 + \omega_1)t \times \frac{\omega_2 - \omega_1}{t}$$

or $\quad 2\alpha\theta = \omega_2^2 - \omega_1^2$ (iii)

From equation (ii),

$$\omega_2 = \omega_1 + \alpha t \quad \text{(iv)}$$

Substituting for ω_2 in equation (i),

$$\theta = \tfrac{1}{2}(\omega_1 + \alpha t + \omega_1)\,t$$

or $\quad \theta = \omega_1 t + \tfrac{1}{2}\alpha t^2$ (v)

Equations (i), (iii), (iv), and (v) above are usually rearranged as follows:

$$\theta = \tfrac{1}{2}(\omega_2 + \omega_1)\,t$$

$$\theta = \omega_1 t + \tfrac{1}{2}\alpha t^2$$

$$\omega_2 = \omega_1 + \alpha t$$

$$\omega_2^2 = \omega_1^2 + 2\alpha\theta$$

These equations are known as the equations of uniform angular motion and should be remembered.

Example 1 A flywheel rotating at 210 rev/min is accelerated uniformly for 4 seconds until its speed is 560 rev/min. Determine the uniform angular acceleration and the number of revolutions made by the flywheel in the 4-second period.

From the equations of uniform angular motion,

$$\omega_2 = \omega_1 + \alpha t$$

$\therefore \quad \alpha = (\omega_2 - \omega_1)/t$

and $\quad \theta = \tfrac{1}{2}(\omega_2 + \omega_1)\,t$

where $\quad \omega_1 = \dfrac{210 \text{ rev/min} \times 2\pi \text{ rad/rev}}{60 \text{ s/min}} = 22 \text{ rad/s}$

$$\omega_2 = \frac{560 \text{ rev/min} \times 2\pi \text{ rad/rev}}{60 \text{ s/min}} = 58.6 \text{ rad/s}$$

and $\quad t = 4\text{ s}$

$\therefore \quad \alpha = \dfrac{58.6 \text{ rad/s} - 22 \text{ rad/s}}{4\text{ s}}$

$$= 9.15 \text{ rad/s}^2$$

i.e. the angular acceleration is 9.15 rad/s^2

$$\theta = \tfrac{1}{2}(58.6 \text{ rad/s} + 22 \text{ rad/s}) \times 4\text{ s}$$

$$= 161.2 \text{ rad}$$

$$\therefore \quad \text{number of revolutions} = \frac{161.2 \text{ rad}}{2\pi \text{ rad/rev}}$$

$$= 25.7$$

i.e. the number of revolutions made by the flywheel in 4 seconds is 25.7.

Example 2 A flywheel is accelerated uniformly from rest for 3 seconds until it is rotating at 210 rev/min. It revolves at 210 rev/min for 4 minutes and is then decelerated uniformly at the rate of 5 rad/s² to rest. For the flywheel, determine (a) the uniform angular acceleration, (b) the deceleration time, (c) the total number of complete revolutions made.

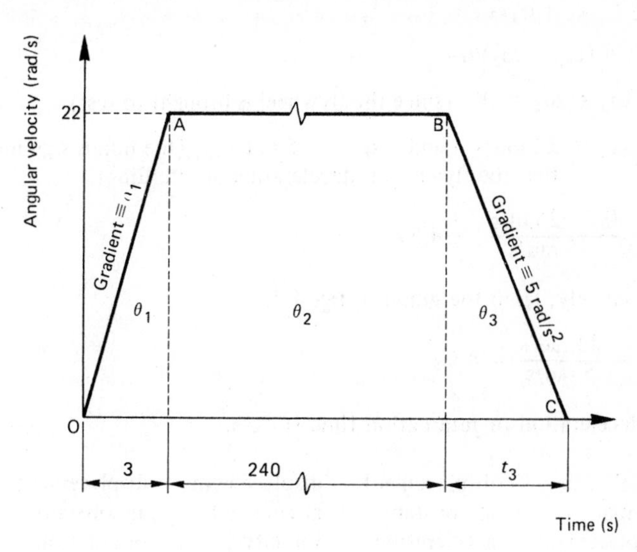

Fig. 4.2

Figure 4.2 illustrates the problem graphically.

a) From the equations of uniform angular motion,

$$\alpha_1 = (\omega_2 - \omega_1)/t_1$$

where $\omega_1 = 0$ (since the flywheel is initially at rest)

$$\omega_2 = \frac{210 \text{ rev/min} \times 2\pi \text{ rad/rev}}{60 \text{ s/min}} = 22 \text{ rad/s}$$

and $t_1 = 3$ s

$$\therefore \quad \alpha_1 = \frac{22 \text{ rad/s} - 0}{3 \text{ s}}$$

$$= 7.33 \text{ rad/s}^2$$

Alternatively, from the graph in fig. 4.2,

$$\alpha_1 = \text{slope of line OB}$$

$$= \frac{22 \text{ rad/s}}{3 \text{ s}}$$

$$= 7.33 \text{ rad/s}^2$$

i.e. the uniform angular acceleration is 7.33 rad/s^2.

b) For the third phase,

$$\omega_2 = \omega_3 + \alpha_3 t_3$$

$$\therefore \quad t_3 = (\omega_3 - \omega_2)/\alpha_3$$

where $\omega_3 = \omega_1 = 0$ (since the flywheel is brought to rest)

$\omega_2 = 22 \text{ rad/s}$ and $\alpha_3 = -5 \text{ rad/s}^2$ (the minus sign indicating that the flywheel is decelerating or retarding)

$$\therefore \quad t_3 = \frac{0 - 22 \text{ rad/s}}{-5 \text{ rad/s}} = 4.5 \text{ s}$$

Alternatively, from the graph in fig. 4.2,

$$t_3 = \frac{22 \text{ rad/s}}{5 \text{ rad/s}} = 4.5 \text{ s}$$

i.e. the deceleration or retardation time is 4.5 s.

c) Total angular displacement = displacement at constant acceleration + displacement at constant velocity + displacement at constant retardation

i.e. $\theta = \theta_1 + \theta_2 + \theta_3$

$$= \tfrac{1}{2}(\omega_2 + \omega_1)t_1 + \omega_2 t_2 + \tfrac{1}{2}(\omega_3 + \omega_2)t_3$$

where $t_2 = 4 \text{ minutes} = 240 \text{ s}$

$$\therefore \quad \theta = [\tfrac{1}{2}(22 \text{ rad/s} + 0) \times 3 \text{ s}] + [22 \text{ rad/s} \times 240 \text{ s}]$$
$$+ [\tfrac{1}{2}(0 + 22 \text{ rad/s}) \times 4.5 \text{ s}]$$

$$= 33 \text{ rad} + 5280 \text{ rad} + 49.5 \text{ rad}$$

$$= 5362.5 \text{ rad}$$

Alternatively, from the graph in fig. 4.2,

$$\text{total angular displacement} = \text{area OABC}$$

$$= \tfrac{1}{2}(3 \text{ s} + 2 \times 240 \text{ s} + 4.5 \text{ s}) \times 22 \text{ rad/s}$$

$$= 5362.5 \text{ rad}$$

$$\text{Total revolutions made by the flywheel} = \frac{5362.5 \text{ rad}}{2\pi \text{ rad/rev}}$$

$$= 853.5$$

i.e. the total number of *complete* revolutions made by the flywheel is 853.

4.2 Torque and angular acceleration

The elemental mass δm at radius r in the disc shown in fig. 4.3 has an angular velocity ω and angular acceleration α. At the instant shown, it also has a *tangential* or *linear* acceleration a.

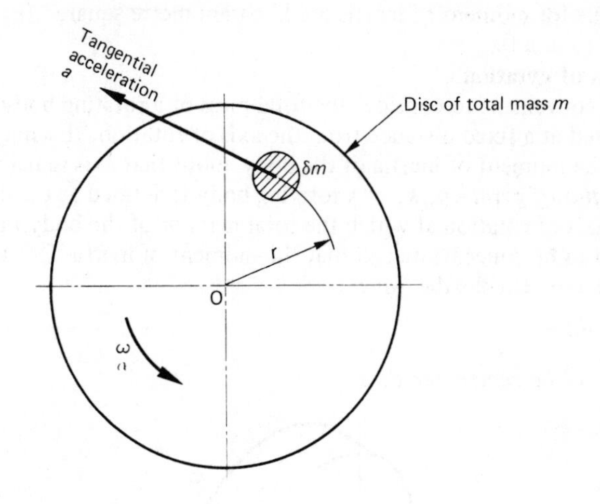

Fig. 4.3 Rotating disc

From Newton's second law of motion, the force acting on δm to produce the acceleration a is given by

$$\delta F = \delta m a$$

and the torque about the centre of rotation O is given by

$$\delta T = \delta F r$$

or $$\delta T = \delta m a r \qquad \text{(vi)}$$

From the relationship between linear and angular motion,

$$a = r\alpha$$

Substituting for a in equation (vi) gives

$$\delta T = \delta m (r\alpha) r$$

$$= \delta m r^2 \alpha$$

The total torque to accelerate the complete disc is given by

$$T = \Sigma \delta T$$

$$= \Sigma \delta m r^2 \alpha$$

The term $\Sigma \delta m r^2$ is called the *mass moment of inertia* or simply the *moment of inertia*, I, of the body about its centre of rotation,

i.e. torque = moment of inertia × angular acceleration

or $T = I\alpha$

which should be remembered.

The units for moment of inertia are kilogram metre squared ($kg\,m^2$).

4.3 Radius of gyration

It is often convenient to consider the total mass of a rotating body as being concentrated at a fixed distance from the axis of rotation; this may be done provided the moment of inertia of the body about that axis remains the same.

The *radius of gyration*, k, of a rotating body is defined as the distance from the axis of rotation at which the total mass m of the body may be considered to be concentrated so that the moment of inertia I of the body about that axis remains the same,

i.e. $I = mk^2$

which should be remembered.

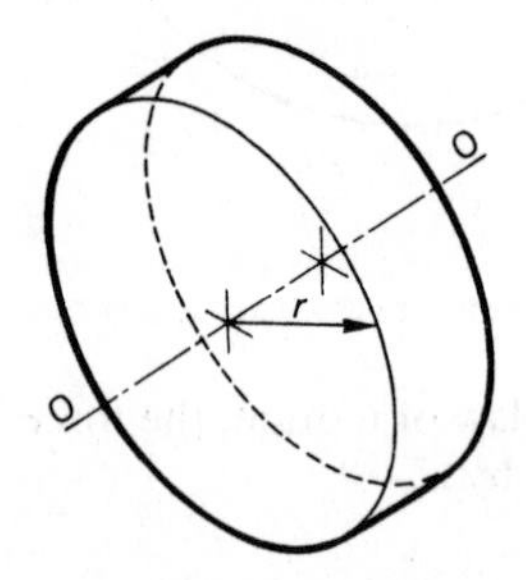

Fig. 4.4

For the solid disc shown in fig. 4.4, the radius of gyration about the axis OO is

$$k = \frac{r}{\sqrt{2}} = \frac{d}{2\sqrt{2}}$$

or $k^2 = \frac{r^2}{2} = \frac{d^2}{8}$

which it is useful to remember.

Thus, if the total mass of the solid disc in fig. 4.4 is concentrated at radius $r/\sqrt{2}$, then the moment of inertia I about the axis OO will remain the same.

Example 1 The rotating parts of an engine have a mass of 2.4 kg and radius of gyration 100 mm. Determine the torque required to overcome the inertia of the rotating parts when the angular acceleration is 15 rad/s².

$$T = I\alpha = mk^2\alpha$$

where $m = 2.4\,\text{kg}$ $k = 100\,\text{mm} = 0.1\,\text{m}$ and $\alpha = 15\,\text{rad/s}^2$

$\therefore$ $T = 2.4\,\text{kg} \times (0.1\,\text{m})^2 \times 15\,\text{rad/s}^2$

$= 0.36\,\text{Nm}$ (since $1\,\text{kg m}^2/\text{s}^2 = 1\,\text{Nm}$)

i.e. the applied torque is 0.36 Nm.

Example 2 An experimental solid-disc flywheel has a mass of 120 kg and an outside diameter of 300 mm. With the flywheel at rest, a constant torque of 10 Nm is applied for a period of 2 seconds. Ignoring the effect of friction, determine for the flywheel (a) the moment of inertia, (b) the angular acceleration, (c) the angular velocity after 2 seconds.

a) $I = mk^2$

where $m = 120\,\text{kg}$ and $k^2 = d^2/8 = (0.3\,\text{m})^2/8 = 0.011\,25\,\text{m}^2$

$\therefore$ $I = 120\,\text{kg} \times 0.011\,25\,\text{m}^2$

$= 1.35\,\text{kg m}^2$

i.e. the moment of inertia is 1.35 kg m².

b) $T = I\alpha$

$\therefore$ $\alpha = T/I$

where $T = 10\,\text{Nm}$

$$\therefore \quad \alpha = \frac{10\,\text{Nm}}{1.35\,\text{kg m}^2}$$

$= 7.41\,\text{rad/s}^2$

i.e. the angular acceleration is 7.41 rad/s².

c) From section 4.1,

$$\omega_2 = \omega_1 + \alpha t$$

where $\omega_1 = 0$ (since the flywheel is initially at rest) and $t = 2\,\text{s}$

$\therefore$ $\omega_2 = 0 + (7.41\,\text{rad/s}^2 \times 2\,\text{s})$

$= 14.82\,\text{rad/s}$

i.e. the velocity after 2 seconds is 14.82 rad/s.

Example 3 In an experiment to determine the radius of gyration of an annulus (i.e. a disc with a hole in the centre), the following observations were made: applied torque 6 N m; angular acceleration 9.2 rad/s²; mass of annulus 14.4 kg. Calculate the radius of gyration.

$$T = I\alpha = mk^2\alpha$$

$$\therefore \quad k = \sqrt{\frac{T}{m\alpha}}$$

where $T = 6\,\text{N m}$ $\quad m = 14.4\,\text{kg}$ $\quad$ and $\quad \alpha = 9.2\,\text{rad/s}^2$

$$\therefore \quad k = \sqrt{\frac{6\,\text{N m}}{14.4\,\text{kg} \times 9.2\,\text{rad/s}^2}}$$

$$= \sqrt{0.0453\,\text{m}^2}$$

$$= 0.213\,\text{m} \quad \text{or} \quad 213\,\text{mm}$$

i.e. the radius of gyration is 213 mm.

4.4 Centripetal acceleration

The disc of radius r shown in fig. 4.5(a) is rotating with uniform angular velocity ω. Relative to the centre of rotation O, the point A has a linear velocity v in a direction perpendicular to OA. This velocity is represented by the vector ***oa*** in fig. 4.5(b). In time δt, A will move through distance δs to

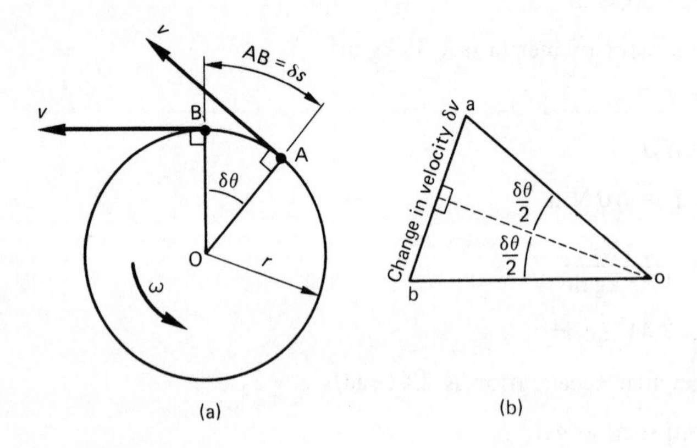

Fig. 4.5 Centripetal acceleration

B. Since the disc is rotating with uniform angular velocity, the linear velocity of B relative to the centre of rotation will also be v. This velocity is represented by the vector ***ob*** in fig. 4.5(b). In fig. 4.5(b), the vector ***ab*** represents the *change* in velocity δv between A and B,

i.e. $\quad$ change in velocity, $\delta v = \boldsymbol{ab} = 2\,\boldsymbol{oa} \sin \dfrac{\delta\theta}{2}$

But $\delta\theta$ is very small,

$$\therefore \quad \sin\frac{\delta\theta}{2} \approx \frac{\delta\theta}{2} \text{ radians}$$

$$\therefore \quad \boldsymbol{ab} = 2\,\boldsymbol{oa}\,\frac{\delta\theta}{2} = \boldsymbol{oa}\,\delta\theta$$

or $\quad \delta v = v\delta\theta$

(since $\boldsymbol{oa} = \boldsymbol{ob} = v$)

For there to be a change in velocity there must be an acceleration of magnitude

$$a = \frac{\delta v}{\delta t}$$

$$= v\frac{\delta\theta}{\delta t}$$

(since the change takes place in time δt)

But $\quad \dfrac{\delta\theta}{\delta t} = \omega$

$$\therefore \quad a = v\omega$$

From the relationship between linear and angular motion,

$$v = r\omega$$

$$\therefore \quad a = \omega^2 r = \frac{v^2}{r}$$

The direction of this acceleration is from a towards b in fig. 4.5(b).

Referring to fig. 4.5(b), as $\delta\theta$ approaches zero, the angle oab approaches 90°, thus the direction of the *instantaneous* change in the velocity of A is *towards* the centre of rotation O. For this reason, the acceleration is termed *centripetal acceleration*; i.e. *bodies which move in a circular path have centripetal acceleration, or acceleration towards the centre of rotation, of magnitude*

$$a = \omega^2 r = \frac{v^2}{r}$$

which should be remembered.

4.5 Centripetal and centrifugal force

If the point A on the rim of the rotating disc in fig. 4.5(a) has a mass m, then, from Newton's second law of motion, a force of magnitude

$$F = m\omega^2 r = m\frac{v^2}{r}$$

will act *on* the mass in a direction towards the centre of rotation. This force is known as the *centripetal force*,

i.e. $$\text{centripetal force} = m\omega^2 r = m\frac{v^2}{r}$$

From Newton's third law of motion, if a force acts *on* the mass, the mass will exert a reaction force which is equal in magnitude but opposite in direction to the applied force. If the applied force causes the mass to accelerate, the reaction force is known as the *inertia force*. For the special case of a mass moving in a circular path, the reaction or inertia force is called the *centrifugal force*; i.e. the centrifugal force has the same magnitude as the centripetal force but its direction is *outwards* from the centre of rotation. Thus

$$\text{centrifugal force} = m\omega^2 r = m\frac{v^2}{r}$$

Typical examples of centripetal and centrifugal forces are

a) the hammer thrower in fig. 4.6 is exerting a *centripetal* force on the hammer head by *pulling* on the wire; the hammer head is trying to move outwards, thus exerting a *centrifugal* pull on the wire;

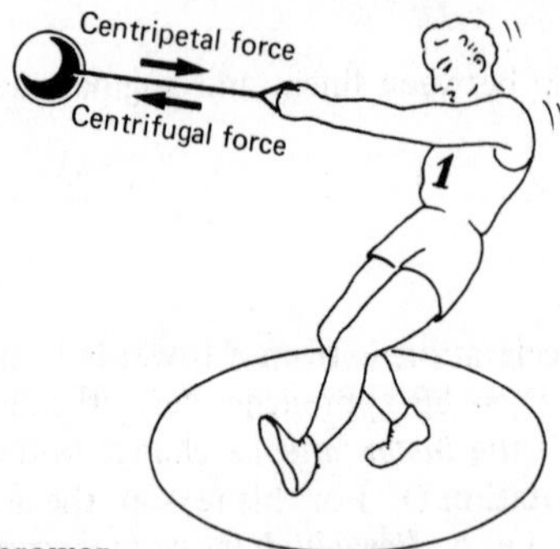

Fig. 4.6 Hammer thrower

b) the 'wall-of-death' motor-cyclist at the fun-fair is prevented from moving outwards because the wall exerts a *centripetal* force on the machine; the machine is applying a *centrifugal* force to the wall;

c) as a train rounds a curve, the outer rail exerts a *centripetal* force on the train while the train exerts a *centrifugal* force on the rail.

Thus centripetal forces act *inwards on the mass* and centrifugal forces act *outwards*. Both forces are of magnitude

$$F = m\omega^2 r = m\frac{v^2}{r}$$

which should be remembered.

Example 1 A casting on a lathe faceplate has a mass of 20 kg which can be considered to be concentrated at a radius of 220 mm. Determine the radial force acting on the spindle bearings when the faceplate is rotating at 90 rev/min. What is the name of this radial force?

Since the casting on the faceplate is rotating, it will exert a *centrifugal* force on the spindle bearings; i.e. the name of the radial force acting on the spindle bearings is *centrifugal force*. Note that the bearings are exerting a *centripetal* force *on* the casting.

Centrifugal force $F = m\omega^2 r$

where $m = 20\,\text{kg}$ $\qquad \omega = \dfrac{90\,\text{rev/min} \times 2\pi\,\text{rad/rev}}{60\,\text{s/min}} = 9.42\,\text{rad/s}$

and $r = 220\,\text{mm} = 0.22\,\text{m}$

$\therefore \quad F = 20\,\text{kg} \times (9.42\,\text{rad/s})^2 \times 0.22\,\text{m}$

$= 390.4\,\text{N}$

i.e. the radial or centrifugal force on the spindle is 390.4 N.

Example 2 A vehicle of mass 900 kg negotiates a curve of mean radius 40 m at a constant speed of 40 km/h. If the track surface is level, determine the centripetal force being exerted on the vehicle. If the coefficient of friction between the tyres and the track surface is 0.6, what will be the maximum speed at which the vehicle can round the curve without sideways slip?

Centripetal force $F = mv^2/r$

where $m = 900\,\text{kg}$ $\qquad v = \dfrac{40\,\text{km/h} \times 1000\,\text{m/km}}{3600\,\text{s/h}} = 11.1\,\text{m/s}$

and $r = 40\,\text{m}$

$\therefore \quad F = \dfrac{900\,\text{kg} \times (11.1\,\text{m/s})^2}{40\,\text{m}} = 2772\,\text{N}$

i.e. the centripetal force is 2772 N and acts towards the centre of the curve.

The maximum centripetal force which can be exerted on the vehicle is equal to the limiting frictional force between the tyres and the track surface. If N is the normal reaction, then

limiting frictional force $= \mu N = \mu m g$

where $\mu = 0.6$ and $g = 9.81\,\text{m/s}^2$

$\therefore$ limiting frictional force $= 0.6 \times 900\,\text{kg} \times 9.81\,\text{m/s}^2$

$= 5297\,\text{N}$

$=$ maximum centripetal force

But centripetal force $F = mv^2/r$

$\therefore$ maximum speed $v = \sqrt{\dfrac{Fr}{m}}$

$$= \sqrt{\frac{5297 \text{ N} \times 40 \text{ m}}{900 \text{ kg}}}$$

$$= 15.34 \text{ m/s} \quad \text{or} \quad 55.2 \text{ km/h}$$

i.e. the maximum speed at which the vehicle can round the curve without sideways slip is 55.2 km/h. This maximum speed is also knows as the *limiting speed*.

Example 3 The centre of gravity of a motor cycle and its rider is 0.7 m above the road surface with the machine in an upright position. Assuming there to be no sideways slip and that the road surface is level, determine the magnitude of the heel-over angle, measured from the vertical, that would be necessary for the motor-cyclist safely to negotiate a curve of radius 50 m at a constant speed of 60 km/h.

As the motor-cyclist rounds the curve, the overturning effect of the *centrifugal* force is resisted by the rider leaning in towards the centre of the curve. This produces a moment about the wheel–road interface at A as shown in fig. 4.7.

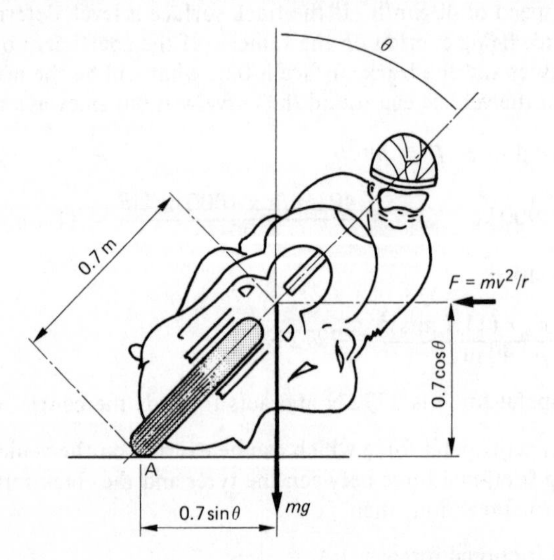

Fig. 4.7

Referring to fig. 4.7, let m be the combined mass of the rider and machine and take moments about A:

$$\Sigma \text{ clockwise moments} = \Sigma \text{ anticlockwise moments}$$

$$\therefore \quad F \times 0.7 \cos\theta = mg \times 0.7 \sin\theta$$

or $$\frac{\sin\theta}{\cos\theta} = \frac{F}{mg}$$

But $\dfrac{\sin\theta}{\cos\theta} = \tan\theta$ and $F = \dfrac{mv^2}{r}$

$$\therefore \quad \tan\theta = \frac{mv^2/r}{mg} = \frac{v^2}{rg}$$

where $v = \dfrac{60\ \text{km/h} \times 1000\ \text{m/km}}{3600\ \text{s/h}} = 16.7\ \text{m/s} \qquad r = 50\ \text{m}$

and $g = 9.81\ \text{m/s}^2$

$$\therefore \quad \tan\theta = \frac{(16.7\ \text{m/s})^2}{50\ \text{m} \times 9.81\ \text{m/s}^2}$$

$$= 0.5686$$

$$\therefore \quad \theta = 29.62^\circ$$

i.e. the heel-over angle is 29.62°.

Example 4 A high-sided vehicle has its centre of gravity 1.8 m above the road surface and a wheel-track width of 2.2 m. What is the minimum speed at which the vehicle could round a 35° banked curve (measured from the horizontal) at a mean radius of 40 m?

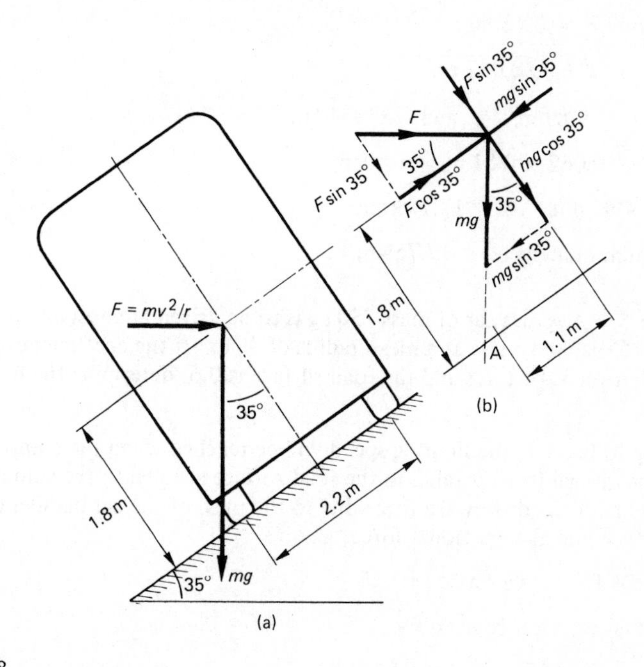

Fig. 4.8

The problem is shown diagrammatically in fig. 4.8(a), where it can be seen that the line of action of the downward force, mg, is *outside* the wheelbase; i.e. the vehicle would topple inwards if it were stationary. Thus, the moment of the *centrifugal* force F about the point A must balance the overturning couple due to the mass of the vehicle. To simplify the calculation, the centrifugal force F and the downward force mg have been resolved into components which are perpendicular and parallel to the road surface, as shown in fig. 4.8(b).

Referring to fig. 4.8(b), the resolved perpendicular components are $F \sin 35°$ and $mg \cos 35°$, which will both produce *clockwise* moments about A. The resolved components parallel to the road are $F \cos 35°$, which will produce a *clockwise* moment about A, and $mg \sin 35°$ which will produce an *anticlockwise* moment about A.

Taking moments about A,

$$\Sigma \text{ clockwise moments} = \Sigma \text{ anticlockwise moments}$$

$$\therefore \quad (1.1 \times F \sin 35°) + (1.1 \times mg \cos 35°) + (1.8 \times F \cos 35°) = (1.8 \times mg \sin 35°)$$

$$0.63\,F + 0.9\,mg + 1.47\,F = 1.03\,mg$$

or

$$2.1\,F = 0.13\,mg$$

But $\quad$ centrifugal force $F = mv^2/r$

$$\therefore \quad 2.1\,mv^2/r = 0.13\,mg$$

or

$$v = \sqrt{0.062\,gr}$$

where $\quad g = 9.81\text{ m/s}^2 \quad$ and $\quad r = 40\text{ m}$

$$\therefore \quad v = \sqrt{(0.062 \times 9.81\text{ m/s}^2 \times 40\text{ m})}$$

$$= 4.93\text{ m/s} \quad \text{or} \quad 17.75\text{ km/h}$$

i.e. the minimum speed is 17.75 km/h.

Example 5 A sports car of mass 750 kg is to be driven at constant speed round a 15° banked curve at a mean radius of 45 m. If the coefficient of friction between the tyres and the road surface is 0.6, determine the limiting speed.

Referring to fig. 4.9, the limiting speed will be reached when the component of the centrifugal force parallel to the road surface is equal to the sum of the components of the downward force due to the mass of the car parallel to the road surface and the frictional force,

i.e. $\quad F \cos 15° = mg \sin 15° + \mu N$

The normal reaction is given by

$$N = F \sin 15° + mg \cos 15°$$

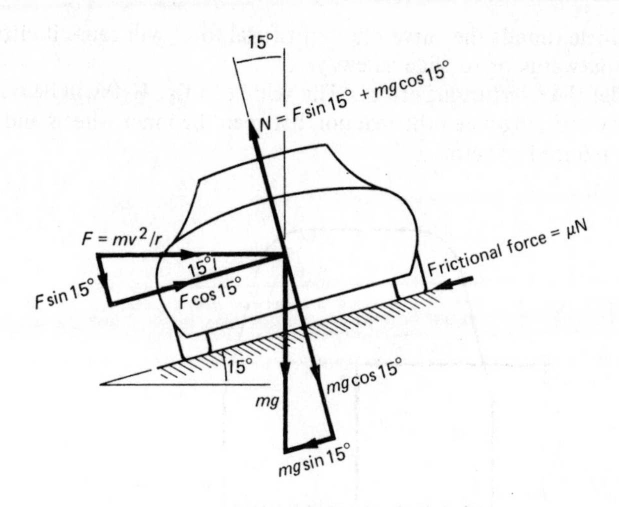

Fig. 4.9

$\therefore \quad F \cos 15^\circ = mg \sin 15^\circ + \mu\,(F \sin 15^\circ + mg \cos 15^\circ)$

where $m = 750\ \text{kg} \qquad g = 9.81\ \text{m/s}^2 \quad$ and $\quad \mu = 0.6$

$\therefore \quad 0.9659F = [750\ \text{kg} \times 9.81\ \text{m/s}^2 \times 0.2588]$

$\qquad\qquad + 0.6\ [0.2588F + (750\ \text{kg} \times 9.81\ \text{m/s}^2 \times 0.9659)]$

$0.9659\,F = 1904\ \text{N} + 0.1553\,F + 4264\ \text{N}$

$\therefore \quad 0.8106\,F = 6168\ \text{N}$

or $\qquad F = 7609\ \text{N}$

But centrifugal force $F = mv^2/r$

where $r = 45\ \text{m}$

$$\therefore \quad 750\ \text{kg} \times \frac{v^2}{45\ \text{m}} = 7609\ \text{N}$$

$$\therefore \quad v = \sqrt{\frac{7609\ \text{N} \times 45\ \text{m}}{750\ \text{kg}}}$$

$$= 21.37\ \text{m/s} \quad \text{or} \quad 76.93\ \text{km/h}$$

i.e. the limiting speed is 76.93 km/h.

Example 6 A vehicle with a wheel-track width of 1.5 m and centre of gravity 1.2 m above the road surface is driven at constant speed round a level curve of mean radius 65 m. If the coefficient of friction between the road surface and the tyres is 0.65, calculate the maximum speed at which the vehicle can safely negotiate the curve.

As the vehicle rounds the curve, the centrifugal force will cause it either to overturn outwards or to slide sideways.

Consider the overturning effect. The vehicle in fig. 4.10 will be on the point of overturning when the reaction between the inner wheels and the road surface is reduced to zero.

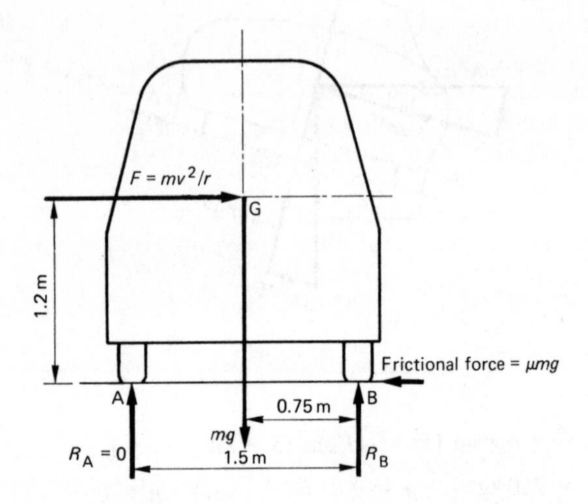

Fig. 4.10

Referring to fig. 4.10, let m be the mass of the vehicle and take moments about the outer wheels B:

$$\Sigma \text{ clockwise moments} = \Sigma \text{ anticlockwise moments}$$

$$\therefore \quad 1.5\,R_A + 1.2\,F = 0.75\,mg$$

But $R_A = 0$ and centrifugal force $F = mv^2/r$

$$\therefore \quad 1.2\,mv^2/r = 0.75\,mg$$

or $$v = \sqrt{0.625\,gr}$$

where $g = 9.81\text{ m/s}^2$ and $r = 65\text{ m}$

$$\therefore \quad v = \sqrt{(0.625 \times 9.81\text{ m/s}^2 \times 65\text{ m})}$$

$$= 19.96\text{ m/s} \quad \text{or} \quad 71.86\text{ km/h}$$

Consider the sliding effect. The vehicle will be on the point of sliding sideways when the centrifugal force is equal to the limiting frictional force between the tyres and the road surface, i.e. when

$$mv^2/r = \mu\,mg$$

or $$v = \sqrt{\mu gr}$$

where $\mu = 0.65$

$\therefore \quad v = \sqrt{(0.65 \times 9.81\ \text{m/s}^2 \times 65\ \text{m})}$

$= 20.36\ \text{m/s} \quad \text{or} \quad 73.3\ \text{km/h}$

i.e. the vehicle would overturn before it would slide, therefore the maximum speed is 71.86 km/h.

Exercises on chapter 4

1 The speed of a flywheel is increased uniformly from 210 rev/min to 250 rev/min in 5 seconds. Determine the angular acceleration and the number of revolutions made by the wheel in the 5-second period.
[0.84 rad/s^2; 19.2 revs]

2 A grinding wheel is rotating at 2800 rev/min when the power is switched off. If it takes 4 minutes for the wheel to come to rest, calculate the angular retardation, assuming it to be uniform. [1.22 rad/s^2]

3 A lift winding drum accelerates uniformly from rest for 1.5 seconds until it is rotating at 80 rev/min. At this speed, the drum makes 2.6 revolutions before coming uniformly to rest in 1.2 seconds. Determine (a) the angular acceleration and retardation, (b) the total time the drum is rotating, (c) the total number of revolutions made by the drum.
[5.6 rad/s^2; 7 rad/s^2; 4.65 s; 4.4 revs]

4 The flywheel of an internal-combustion engine is accelerated uniformly at 15 rad/s^2 for 3 seconds, until the engine speed is 3500 rev/min. What is the initial speed? If the rotating parts of the engine have a moment of inertia of 1.3 kg m^2, determine the accelerating torque. [3070 rev/min; 19.5 N m]

5 Define 'radius of gyration'.
The mass of a flywheel can be assumed to be concentrated at a radius of 215 mm. If the torque required to give the flywheel an angular acceleration of 6 rad/s^2 is 11 N m, estimate the mass of the flywheel. [39.7 kg]

6 An experimental solid-disc flywheel is 300 mm diameter and has a mass of 70 kg. Experiments show that, at all speeds of rotation, the frictional resistance to motion is equivalent to a torque of 0.15 N m. If when the flywheel is rotating at 50 rev/min the driving power is removed, calculate (a) the angular retardation, (b) the time taken for the flywheel to come to rest, (c) the total number of revolutions made by the flywheel during retardation. [0.19 rad/s^2; 27.6 s; 11.5 revs]

7 A car wheel and tyre, having a mass of 35 kg and a rolling diameter of 0.61 m, is balanced with the wheel rotating at an equivalent road speed of 112 km/h. If a torque of 20 N m is required to accelerate the wheel uniformly to its running speed in 5 seconds, calculate the radius of gyration of the wheel and tyre. [0.167 m]

8 A mass of 60 g is required to balance a motor-car wheel. If the mass is attached to the 0.33 m diameter rim, calculate the centrifugal force acting on the rim when the wheel is rotating at 470 rev/min. [24 N]

9 The pick-up arm on a record player has an effective mass of 80 g at all positions on the turntable. What is the centrifugal force acting on a record playing at $33\frac{1}{3}$ rev/min when the stylus is at a radius of 145 mm? Will the force increase or decrease as the stylus moves towards the centre of the record? [0.14 N; decrease]

10 When negotiating a curve on a level road, a cyclist is limited to a heel-over angle of 15° measured from the vertical. Determine the maximum speed at which the cyclist can safely round a curve of 40 m radius. [36.9 km/h]

11 The coefficient of friction between the tyres of a motor cycle and the vertical wall of the 'wall of death' at the fun-fair is 0.6. Calculate the minimum speed the rider must attain before moving on to the vertical wall where the diameter is 20 m. If the combined mass of rider and machine is 370 kg, what is the centrifugal force acting on the wall? [46 km/h; 6.05 kN]

12 A vehicle with a track width of 1.6 m and centre of gravity 1.4 m above the road surface rounds a level curve of mean radius 60 m. If the coefficient of friction between the tyres and the road surface is 0.65, determine the limiting speed. [66 km/h]

13 If the vehicle in question 12 rounds a 10° banked curve of mean radius 50 m, determine the limiting speed. [84.6 km/h]

14 If the stall speed of an aircraft is 150 km/h, what is the minimum diameter at which it can 'loop the loop' (i.e. turn in a circle whose axis is parallel with the ground)? When turning in this circle, determine the force acting on the aircraft when it is at the bottom of the loop, if the mass of the aircraft is 1200 kg. [354 m; 23.54 kN]

15 A disc has a mass of 1.4 tonne (1 tonne = 1000 kg) and is accelerated from rest at the rate of 2 rad/s^2. If the diameter of the disc is 1.5 m, calculate (a) the radius of gyration of the disc, (b) the accelerating torque, (c) the speed of the disc, in rev/min, after 10 seconds, (d) the total number of revolutions made by the disc in 10 seconds.
[0.53 m; 787.5 N m; 191 rev/min; 15.9 rev]

16 A point on the periphery of a rotating disc has an instantaneous linear velocity of 42 m/s. If the diameter of the disc is 350 mm, calculate
(a) the angular velocity of the disc, (b) the angular acceleration of the disc if, after 20 seconds, the instantaneous linear velocity of the same point has increased uniformly to 85 m/s. [240 rad/s; 12.29 rad/s^2]

17 A vehicle having a mass of 4 tonne is travelling round a level curve of mean radius 20 m at a speed of 20 km/h. If the track width is 2 m and the surface of the vehicle is taken as equivalent to a rectangle 4.5 m high and 6.5 m long with the centre of gravity 1.5 m above the road surface, find the least normal wind pressure which would overturn the vehicle. [683.3 N/m^2]

5 Kinetic energy

5.1 Introduction

A body which is in motion possesses *kinetic energy*. The amount of kinetic energy in the body depends upon the *mass* and the *velocity* of the body.

There are two forms of kinetic energy:

i) *translatory* or *linear*;
ii) *rotational* or *angular*.

5.2 Translatory or linear kinetic energy

Consider the stationary body of mass m shown in fig. 5.1(a) at position A. When acted upon by the force F, the body moves with uniform acceleration a to position B in time t. At B, the body has a velocity v. The motion of the body is illustrated in the velocity–time graph, fig. 5.1(b).

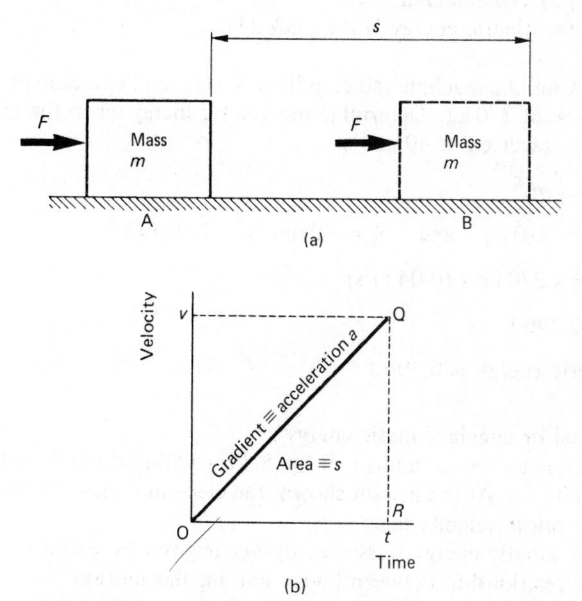

Fig. 5.1 Translatory or linear kinetic energy

Work done on the body = force x distance moved by the body

i.e. $W = Fs$

From Newton's second law of motion,

force F = mass x acceleration

i.e. $F = ma$

$\therefore$ $W = mas$

From the velocity-time graph, fig. 5.1(b),

acceleration $a \equiv$ gradient or slope of the line OQ $= \dfrac{v}{t}$

and distance travelled $s \equiv$ area OQR $= \frac{1}{2}vt$

$$\therefore \quad W = m\frac{v}{t}\ \tfrac{1}{2}vt$$

$$= \tfrac{1}{2}mv^2$$

At position B, the body possesses kinetic energy, the amount being equal to the work done on the body,

i.e. kinetic energy (k.e.) = work W done on the body

or k.e. $= \frac{1}{2}mv^2$

which should be remembered.

The unit for kinetic energy is the *joule* (J).

Example A milling-machine table, milling fixture, and component have a combined mass of 370 kg. Determine the kinetic energy when the table is being rapidly traversed at 40 mm/s.

k.e. $= \frac{1}{2}mv^2$

where m = 370 kg and v = 40 mm/s = 0.04 m/s

$\therefore$ k.e. $= \frac{1}{2} \times 370\text{ kg} \times (0.04\text{ m/s})^2$

$= 0.296\text{ J}$

i.e. the kinetic energy is 0.296 J.

5.3 Rotational or angular kinetic energy

The elemental mass δm at radius r in fig. 5.2 is rotating about O with an angular velocity ω. At the instant shown, the elemental mass also has a *tangential* or *linear* velocity v.

The *linear* kinetic energy possessed by δm is given by $\frac{1}{2}\delta mv^2$.

From the relationship between linear and angular motion,

$$v = r\omega$$

$\therefore$ kinetic energy $= \frac{1}{2}\delta m\,(r\omega)^2$

and total kinetic energy $= \Sigma\frac{1}{2}\delta mr^2\omega^2$

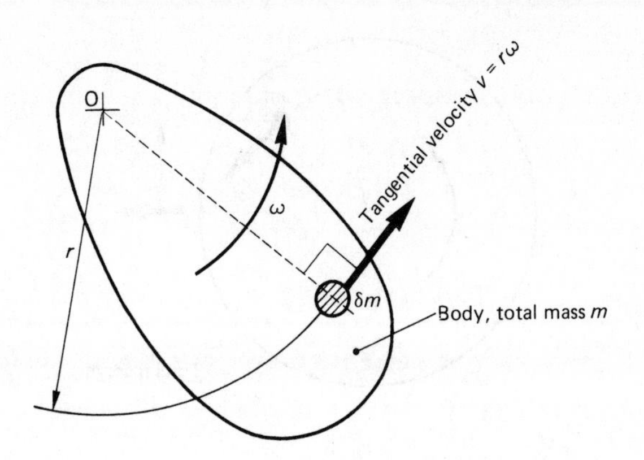

Fig. 5.2 Rotational or angular kinetic energy

From section 4.2,

$$\Sigma\, \delta m r^2 = I \quad \text{(the moment of inertia of the body about O)}$$

and $$I = mk^2$$

where k = radius of gyration about O

$\therefore$ total *angular* kinetic energy (k.e.) $= \frac{1}{2}I\omega^2 = \frac{1}{2}mk^2\omega^2$

which should be remembered.

The unit for angular kinetic energy is the *joule* (J).

Example A winding drum has a mass of 950 kg and radius of gyration 0.82 m. Determine the kinetic energy in the drum when it is rotating at 45 rev/min.

$$\text{k.e.} = \frac{1}{2}I\omega^2 = \frac{1}{2}mk^2\omega^2$$

where $m = 950\,\text{kg} \qquad k = 0.82\,\text{m}$

and $$\omega = \frac{45\,\text{rev/min} \times 2\pi\,\text{rad/rev}}{60\,\text{s/min}} = 4.71\,\text{rad/s}$$

$$\therefore \text{k.e.} = \frac{1}{2} \times 950\,\text{kg} \times (0.82\,\text{m})^2 \times (4.71\,\text{rad/s})^2$$

$$= 7085\,\text{J}$$

i.e. the kinetic energy in the drum is 7085 J.

5.4 Total kinetic energy of a body having both linear and angular motion

The body shown in fig. 5.3(a) is rotating about its centre of gravity G with an angular velocity ω, and at the same time has a linear velocity v. (An example of this condition is a rolling wheel.) At the instant shown, the elemental mass

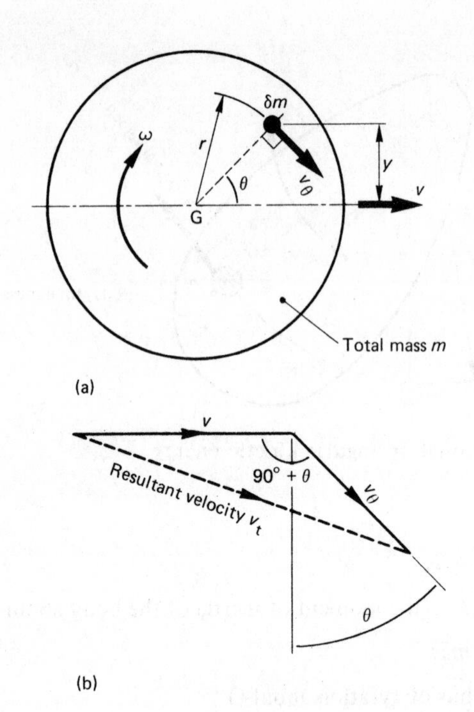

Fig. 5.3 Total kinetic energy

δm at radius r has a *tangential* or *linear* velocity v_θ. The *resultant* or *total* velocity v_t of the elemental mass with respect to the centre of gravity G may be found by vectorial addition as shown in fig. 5.3(b).

From fig. 5.3(b) and using the cosine rule,

$$v_t^2 = v^2 + v_\theta^2 - 2vv_\theta \cos(90^\circ + \theta)$$

But $\cos(90^\circ + \theta) = \sin\theta = y/r$

and, from the relationship between linear and angular motion,

$$v_\theta = r\omega$$

$$\therefore \quad v_t^2 = v^2 + (r\omega)^2 - 2v(r\omega)\frac{y}{r}$$

$$= v^2 + r^2\omega^2 - 2yv\omega$$

The kinetic energy possessed by δm is given by $\frac{1}{2}\delta m v_t^2$

and $\quad$ total kinetic energy $= \Sigma\frac{1}{2}\delta m v_t^2$

$$= \Sigma\tfrac{1}{2}\delta m\,(v^2 + r^2\omega^2 - 2yv\omega)$$

$$= \Sigma\tfrac{1}{2}\delta m v^2 + \Sigma\,\tfrac{1}{2}\delta m r^2\omega^2 - \Sigma\,\delta m y v\omega$$

The term $\delta m y$ is the first moment of the elemental mass about the centre of gravity and, by definition, $\Sigma \delta m y$ about the centre of gravity is zero,

i.e. $\Sigma \delta m y = 0$

$\therefore$ total kinetic energy $= \frac{1}{2} mv^2 + \frac{1}{2} I\omega^2$

i.e. when a body has both linear and angular velocity, the total kinetic energy is the sum of the linear and angular kinetic energies, *which should be remembered.*

Example 1 A solid disc, having an outside diameter of 0.5 m and a mass of 151 kg, rolls without slipping along a level surface with a velocity of 2 m/s. What is the total kinetic energy in the disc?

$$\text{Total k.e.} = \tfrac{1}{2} mv^2 + \tfrac{1}{2} I\omega^2$$
$$= \tfrac{1}{2} m(v_2 + k_2\omega^2)$$

since $I = mk^2$

where $m = 151\text{ kg}$ $\quad v = 2\text{ m/s}$ $\quad \omega = \frac{v}{r} = \frac{2\text{ m/s}}{0.25\text{ m}} = 8\text{ rad/s}$

and, for a solid disc, $k^2 = d^2/8 = (0.5\text{ m})^2/8 = 0.031\,25\text{ m}^2$

$$\therefore \quad \text{total k.e.} = \tfrac{1}{2} \times 151\text{ kg} \times [(2\text{ m/s})^2 + 0.031\,25\text{ m}^2 \times (8\text{ rad/s})^2]$$
$$= 453\text{ J}$$

i.e. the total kinetic energy in the disc is 453 J.

Example 2 The outside diameter and moment of inertia of a lift winding drum are 1.6 m and 130 kg m^2 respectively. The lift cage has a mass of 300 kg. Ignoring the effect of the rope, calculate the total kinetic energy in the system when the cage is being raised at a uniform speed of 1.2 m/s.

$$\text{Total k.e.} = \tfrac{1}{2} mv^2 + \tfrac{1}{2} I\omega^2$$

where $m = 300\text{ kg}$ $\quad v = 1.2\text{ m/s}$ $\quad I = 130\text{ kg m}^2$

and $\omega = \frac{v}{r} = \frac{1.2\text{ m/s}}{0.8\text{ m}} = 1.5\text{ rad/s}$

$$\therefore \quad \text{total k.e.} = [\tfrac{1}{2} \times 300\text{ kg} \times (1.2\text{ m/s})^2] + [\tfrac{1}{2} \times 130\text{ kg m}^2 \times (1.5\text{ rad/s})^2]$$
$$= 216\text{ J} + 146.25\text{ J}$$
$$= 362.25\text{ J}$$

i.e. the total kinetic energy in the system is 362.25 J.

Example 3 The motor in a power press rotates at five times the speed of the driven flywheel. The flywheel is required to provide 40 kJ of energy while its speed is reduced from 210 rev/min to 150 rev/min. Determine the moment of inertia of the flywheel. If the speed of the flywheel is increased uniformly from 150 rev/min to 210 rev/min in 15 seconds and the efficiency of the drive is 75%, determine the torque at the motor shaft.

$$\textit{Change} \text{ in kinetic energy} = \tfrac{1}{2}I(\omega_2^2 - \omega_1^2)$$

$$\therefore \quad I = \frac{2 \times \text{change in k.e.}}{\omega_2^2 - \omega_1^2}$$

where change in k.e. = 40 kJ = 40×10^3 J

$$\omega_1 = \frac{150 \text{ rev/min} \times 2\pi \text{ rad/rev}}{60 \text{ s/min}} = 15.7 \text{ rad/s}$$

and

$$\omega_2 = \frac{210 \text{ rev/min} \times 2\pi \text{ rad/rev}}{60 \text{ s/min}} = 22 \text{ rad/s}$$

$$\therefore \quad I = \frac{2 \times 40 \times 10^3 \text{ J}}{(22 \text{ rad/s})^2 - (15.7 \text{ rad/s})^2}$$

$$= 336.8 \text{ kgm}^2$$

i.e. the moment of inertia is 336.8 kgm².

From section 4.1,

$$\text{angular acceleration } \alpha = (\omega_2 - \omega_1)/t$$

where $t = 15$ s

$$\therefore \quad \alpha = \frac{22 \text{ rad/s} - 15.7 \text{ rad/s}}{15 \text{ s}}$$

$$= 0.42 \text{ rad/s}^2$$

From section 4.2,

$$\text{torque at the flywheel } T_f = I\alpha$$

$$= 336.8 \text{ kgm}^2 \times 0.42 \text{ rad/s}^2$$

$$= 141.5 \text{ Nm}$$

The efficiency of the drive, η (*eta*), is given by

$$\eta = \frac{\text{power output}}{\text{power input}}$$

i.e.

$$\eta = \frac{T_f \omega_f}{T_m \omega_m}$$

$$\therefore \quad \text{torque at motor shaft } T_m = \frac{T_f \omega_f}{\eta \omega_m}$$

where ω_f = speed of flywheel $\quad \omega_m$ = motor speed = $5\,\omega_f$

and η = 75% = 0.75

$$\therefore \quad T_m = \frac{141.5\,\text{Nm} \times \omega_f}{0.75 \times 5\,\omega_f}$$

$$= 37.7\,\text{Nm}$$

i.e. the torque at the motor shaft is 37.7 Nm.

Example 4 An electric motor, which provides a constant output power of 5 kW when rotating at 1440 rev/min, drives a flywheel through a 4:1 reduction gearbox and slipping clutch, the drive being 80% efficient. If it takes 5 seconds for the flywheel to attain its normal running speed, determine its moment of inertia. What will be the kinetic energy in the flywheel at the normal running speed?

$$P = T\omega$$

$$\therefore \quad T = P/\omega$$

where $P = 5\,\text{kW} = 5 \times 10^3\,\text{W}$

and $$\omega_m = \frac{1440\,\text{rev/min} \times 2\pi\,\text{rad/rev}}{60\,\text{s/min}} = 150.8\,\text{rad/s}$$

$$\therefore \quad T_m = \frac{5 \times 10^3\,\text{W}}{150.8\,\text{rad/s}}$$

$$= 33.16\,\text{Nm}$$

$$= \text{torque at the motor shaft}$$

$$\eta = \frac{\text{power output}}{\text{power input}} = \frac{T_f \omega_f}{T_m \omega_m}$$

$$\therefore \quad \text{torque at the flywheel } T_f = \eta T_m \frac{\omega_m}{\omega_f}$$

where $\eta = 0.8$ and $\omega_f = \omega_m/4 = 37.7\,\text{rad/s}$

$$\therefore \quad T_f = 0.8 \times 33.16\,\text{Nm} \times \frac{150.8\,\text{rad/s}}{37.7\,\text{rad/s}}$$

$$= 106.11\,\text{Nm}$$

$$T_f = I\alpha$$

$$\therefore \quad I = T_f/\alpha$$

and $$\alpha = \frac{\omega_2 - \omega_1}{t} = \frac{37.7\,\text{rad/s} - 0}{5\,\text{s}} = 7.54\,\text{rad/s}^2$$

$$\therefore \quad I = \frac{106.11\,\text{Nm}}{7.54\,\text{rad/s}}$$

$$= 14.07\,\text{kgm}^2$$

i.e. the moment of inertia is 14.07 kgm².

$$\text{Angular kinetic energy} = \tfrac{1}{2} I \omega_f^2$$
$$= \tfrac{1}{2} \times 14.07\,\text{kgm}^2 \times (37.7\,\text{rad/s})^2$$
$$= 10\,000\,\text{J} \quad \text{or} \quad 10\,\text{kJ}$$

i.e. the kinetic energy is 10 kJ.

Exercises on chapter 5

1 Calculate the translatory kinetic energy in a vehicle of mass 900 kg travelling with a speed of 60 km/h. [62.75 kJ]

2 A flywheel of mass 150 kg has a radius of gyration of 450 mm and is rotating at 300 rev/min. Calculate the kinetic energy in the flywheel. [15 kJ]

3 The rotating parts of an engine may be considered to have a mass of 3.5 kg acting at a radius of gyration of 0.1 m. If the mass of the engine itself is 80 kg, determine the total kinetic energy in the engine when it is propelling the vehicle at 40 km/h while rotating at 2100 rev/min. [5.775 kJ]

4 A solid disc of mass 15 kg and outside diameter 150 mm is allowed to roll from rest down a plane which is inclined at 6° to the horizontal. If there is no slipping between the disc and the surface, determine the total kinetic energy in the disc when it has travelled 3 m down the incline. What will be its linear velocity and speed of rotation (in rev/min) at this point? (Hint: at the top of the incline, the disc will possess only potential energy.) [46.21 J; 2.027 m/s; 258 rev/min]

5 A lift winding drum is 600 mm diameter and has a moment of inertia of 115 kgm^2, while the lift cage has a mass of 700 kg. Ignoring the effect of the rope, determine the total kinetic energy in the lift system when the drum is rotating at 60 rev/min. [3513.6 J]

6 The motor driving the winding drum in question 5 rotates at 1410 rev/min. If the efficiency of the drive between the drum and the motor shaft is 65%, determine the power at the motor to accelerate the lift system from rest to its running speed of 60 rev/min in 3 seconds. [2333 W]

7 A flywheel of mass 8 tonne (1 tonne = 1000 kg) has a radius of gyration of 1.6 m. Determine the kinetic energy in the flywheel when it is rotating at 90 rev/min. If the speed of the flywheel is reduced to 80 rev/min, what will be the change in kinetic energy. [909.6 kJ; 190.9 kJ]

8 A press flywheel with a radius of gyration of 0.5 m and a mass of 900 kg is rotating at 480 rev/min. Determine the kinetic energy in the flywheel. If 10% of this energy is required during a press operation, determine the percentage reduction in the speed of the flywheel. [284 kJ; 19%]

9 A flywheel stores 500 kJ of kinetic energy when rotating at 250 rev/min. What will be the speed when the kinetic energy is reduced to 400 kJ? [223.5 rev/min]

10 A pump delivers water of density 1000 kg/m^3 through a nozzle of diameter 20 mm with a velocity of 40 m/s. Find the mass of water discharged per second through the nozzle and thus the kinetic energy per second in the water. [12.57 kg/s; 10.056 kJ]

11 A chain connecting two wheels is 4 m long and has a mass of 2 kg per metre length. The smaller driving wheel is 0.4 m diameter and has a moment of inertia of 15 kg m^2, while the moment of inertia of the larger driven wheel is 18 kgm^2. If the speed ratio is 3:2, determine the speed of the driving wheel when the total kinetic energy in the system is 11.51 kJ. What is the linear speed of the chain? [300 rev/min; 6.284 m/s]

12 The power required to accelerate a flywheel uniformly from rest to its running speed of 300 rev/min in 4 seconds is 5 kW. Determine (a) the uniform acceleration, (b) the accelerating torque, (c) the moment of inertia of the flywheel, (d) the kinetic energy in the flywheel when it is rotating at 300 rev/min. [7.855 rad/s^2; 318.3 Nm; 40.52 kgm^2; 20 kJ]

13 The total mass of a waggon is 2.5 tonne, of which the mass of each wheel is 0.25 tonne. Each wheel has a diameter of 0.7 m and a radius of gyration of 0.3 m. Determine the kinetic energy in the waggon when it has a linear velocity of 18 m/s. [524 kJ]

14 A lift cage of mass 750 kg is raised by a cable wrapped around a winding drum 2.4 m in diameter. The cage is accelerated uniformly from rest at the rate of 3 m/s^2 for 5 seconds. Determine the tension in the cable during the acceleration period.

If the winding drum has a mass of 250 kg and its radius of gyration is 0.85 m, ignoring the mass of the cable, calculate (a) the accelerating torque, (b) the total kinetic energy in the lift system after 5 seconds.
[9607.5 N; 11.98 kNm; 98.486 kJ]

15 A solid disc of mass 3.6 kg and outside diameter 150 mm is mounted on a spindle 15 mm in diameter. The spindle is in contact with two smooth parallel rails which are inclined at 7° to the horizontal. Determine the linear velocity and angular velocity of the disc when it has rolled, without slipping, 2 m along the rails. Ignore the effect of the spindle on the inertia of the disc.
[0.306 m/s; 40.8 rad/s]

6 Simple oscillations

6.1 Periodic time and frequency

An oscillation is defined as a to-and-fro motion about a mean position and is completed when the motion of M in fig. 6.1 is from O to A to B and back to O.

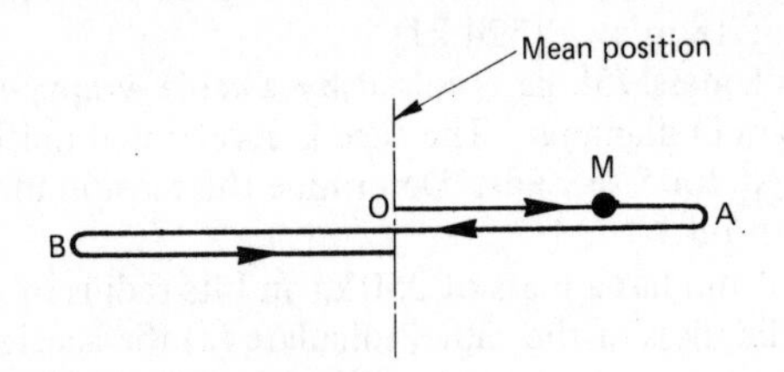

Fig. 6.1 Definition of an oscillation

The time required to complete one cycle of the motion is known as the *period* or *periodic time*, T, of the oscillation, and the number of cycles made per unit time is known as the *frequency*, f, of the oscillation. If the periodic time for each successive oscillation is the same (i.e. T), then

$$\text{frequency} = \frac{1}{\text{periodic time}}$$

or

$$f = \frac{1}{T}$$

which it is useful to remember.

The unit for frequency is the *hertz* (Hz) and for periodic time the *second* (s). It is important to note that

1 Hz = 1 cycle/second or 1 oscillation/second
or 1 revolution/second

6.2 Displacement and amplitude

In oscillatory motion, the *displacement*, x, is measured outwards from either side of the mean position, as shown in fig. 6.2. The maximum displacement on either side of the mean position is known as the *amplitude* of the motion.

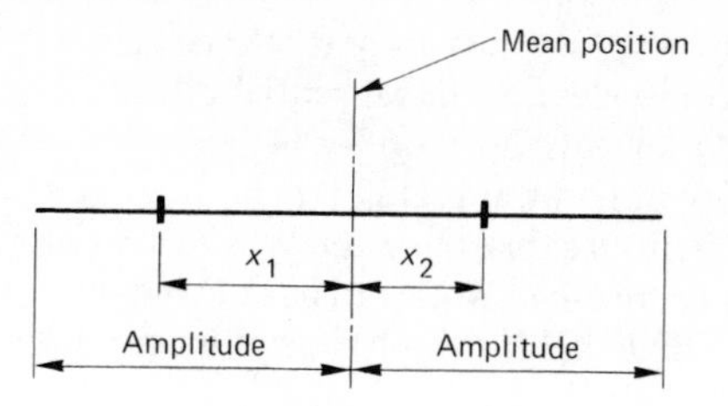

Fig. 6.2 Measurement of displacement and amplitude

6.3 Simple harmonic motion

The most common form of oscillatory motion is *simple harmonic motion* (S.H.M.). An oscillating body will have S.H.M. if, at any instant during the cycle,

i) the acceleration of the body is proportional to its displacement from the mean position, *and*
ii) the acceleration of the body is *always directed towards* the mean position.

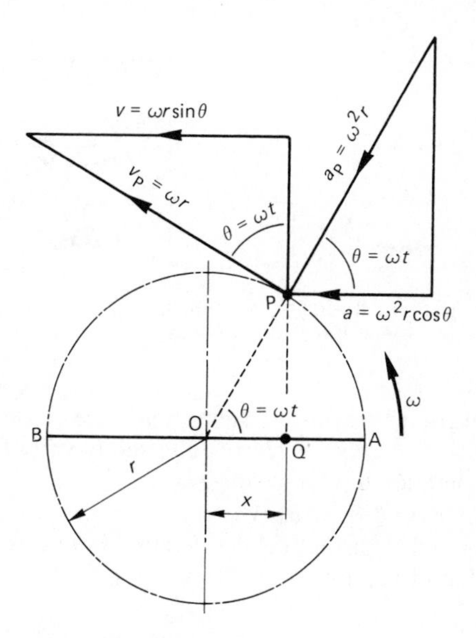

Fig. 6.3 Simple harmonic motion

Consider the motion of Q, which is the projection of P on to the line AB in fig. 6.3. P is rotating in a circle of radius r with uniform angular velocity ω, causing Q to oscillate between A and B about the mean position O. In the position shown, where P has rotated through angle θ in time t, i.e. $\theta = \omega t$,

i) the displacement of Q is $x = r\cos\theta = r\cos\omega t$

ii) the horizontal component of the tangential velocity of P relative to O is the velocity of Q towards O;
i.e. the velocity of Q towards O is $v = -\omega r\sin\theta = -\omega r\sin\omega t$
The minus sign indicates that the velocity is *increasing* while the displacement is *decreasing*. Notice that, as Q moves from O to B, the velocity is *decreasing* and the displacement is *increasing*.

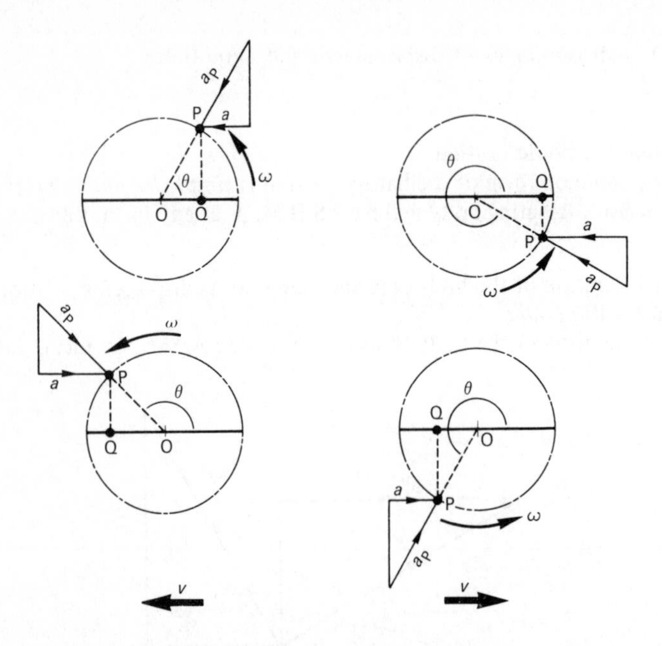

Fig. 6.4 Effect of θ on direction of velocity and acceleration

iii) the horizontal component of the *centripetal* acceleration of P towards O is the acceleration of Q and is *always* directed towards O (see fig. 6.4);
i.e. the acceleration of Q towards O is

$$a = -\omega^2 r\cos\theta = -\omega^2 r\cos\omega t = -\omega^2 x$$

The minus sign indicates that the acceleration is *always* in the opposite direction to the displacement.

Thus, after any time t,

$$\text{displacement } x = r\cos\omega t \qquad \text{(i)}$$

$$\text{velocity} \quad v = \frac{dx}{dt} = -\omega r\sin\omega t \qquad \text{(ii)}$$

$$\text{acceleration} \quad a = \frac{dv}{dt} = -\omega^2 r\cos\omega t = -\omega^2 x \qquad \text{(iii)}$$

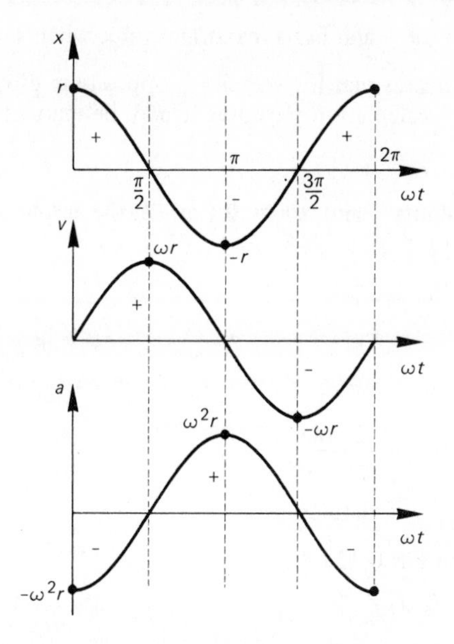

Fig. 6.5 Relationship between displacement, velocity, and acceleration

These equations, *which should be remembered*, are illustrated in the graphs of fig. 6.5.

In equation (iii), the term $-\omega^2$ is constant; therefore

acceleration of Q towards O = constant x displacement from O

i.e. the acceleration of Q is proportional to its displacement from the mean position O and is always directed towards O; thus, Q is oscillating between A and B with *simple harmonic motion*. Notice that the constant is the square of the angular velocity of P.

Q will make one complete oscillation when P has rotated through 2π radians with an angular velocity ω,

$$\therefore \quad \text{periodic time } T = \frac{2\pi}{\omega}$$

which should be remembered.

The frequency of the oscillation of Q is

$$f = \frac{1}{T} = \frac{\omega}{2\pi}$$

which it is useful to remember.

Example 1 The amplitude of motion of a body oscillating with simple harmonic motion is 0.1 m. If the maximum velocity is 4 m/s, determine the periodic time and the frequency of the oscillation.

$$v = -\omega r \sin \omega t \quad \text{and has a maximum value when } \sin \omega t = 1$$

(the minus sign indicates that the velocity is opposite in direction to the displacement – for calculation purposes it may be ignored)

$$\therefore \quad \omega = v_{max.}/r$$

where $v_{max.} = 4$ m/s and $r = 0.1$ m, i.e. the amplitude of the motion

$$\therefore \quad \omega = \frac{4 \text{ m/s}}{0.1 \text{ m}}$$

$$= 40 \text{ rad/s}$$

$$\text{Periodic time } T = 2\pi/\omega$$

$$= \frac{2\pi \text{ rad}}{40 \text{ rad/s}}$$

$$= 0.157 \text{ s}$$

i.e. the periodic time is 0.157 s.

$$\text{Frequency } f = 1/T$$

$$= \frac{1}{0.157 \text{ s}}$$

$$= 6.37 \text{ Hz}$$

i.e. the frequency is 6.37 Hz.

Example 2 A body oscillating with simple harmonic motion has an amplitude of 1.2 m and a periodic time of 5 seconds. Determine the time required for the body to move 0.5 m from its outermost position. What will be the velocity of the body at this point?

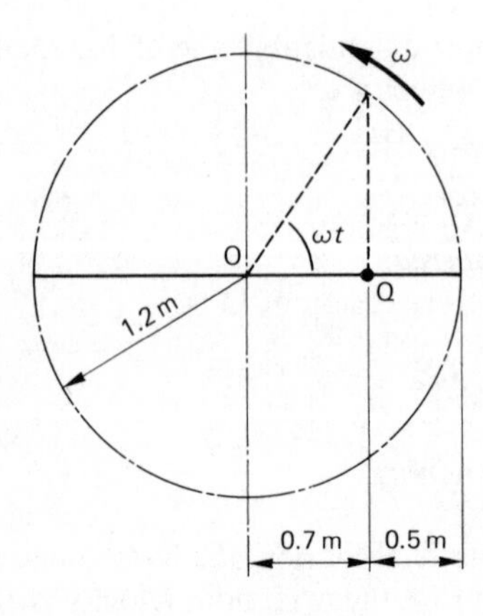

Fig. 6.6

The problem is shown diagrammatically in fig. 6.6.

Periodic time $T = 2\pi/\omega = 5\text{ s}$

$$\therefore \quad \omega = \frac{2\pi\text{ rad}}{5\text{ s}}$$

$$= 1.26\text{ rad/s}$$

From fig. 6.6,

$$\cos \omega t = \frac{0.7\text{ m}}{1.2\text{ m}} = 0.5833$$

$$\therefore \quad \omega t = 54.32° = 0.948\text{ rad}$$

$$\therefore \quad t = \frac{0.948\text{ rad}}{1.26\text{ rad/s}}$$

$$= 0.75\text{ s}$$

i.e. the time is 0.75 s.

$$v = \omega r \sin \omega t \quad \text{(ignoring the sign)}$$

$$= 1.26\text{ rad/s} \times 1.2\text{ m} \times \sin 54.32°$$

$$= 1.228\text{ m/s}$$

i.e. the velocity is 1.228 m/s.

Example 3 The total movement of a body oscillating with simple harmonic motion is 150 mm. If the frequency of the oscillation is 8 Hz, determine the maximum velocity and the maximum acceleration of the body.

Since the total movement is 150 mm, the amplitude of the motion is 75 mm or 0.075 m, as shown in fig. 6.7.

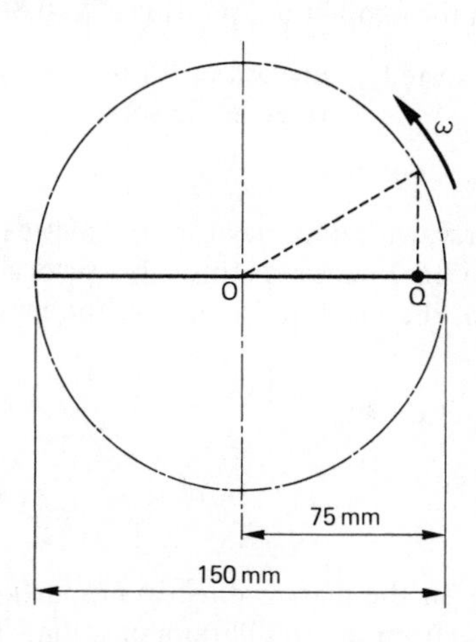

Fig. 6.7

Frequency $f = \omega/2\pi = 8\text{ Hz}$

$$\therefore \quad \omega = 2\pi \text{ rad} \times 8\text{ Hz}$$

$$= 50.27\text{ rad/s}$$

$v = \omega r \sin \omega t$ (ignoring the sign) and has a maximum value when $\sin \omega t = 1$

and $r = 0.075\text{ m}$

$$\therefore \quad v_{max.} = 50.27\text{ rad/s} \times 0.075\text{ m}$$

$$= 3.77\text{ m/s}$$

i.e. the maximum velocity is 3.77 m/s.

$a = -\omega^2 r \cos \omega t$ and has a maximum value when $\cos \omega t = 1$

(the minus sign indicates that the acceleration is always in the opposite direction to the displacement – for calculation purposes it may be ignored)

$$\therefore \quad a_{max.} = \omega^2 r$$

$$= (50.27\text{ rad/s})^2 \times 0.075\text{ m}$$

$$= 189.5\text{ m/s}^2$$

i.e. the maximum acceleration is 189.5 m/s² and is towards the mean position.

6.4 Free linear oscillations of a mass on a vertical spring

The spring shown in fig. 6.8(a) is of negligible mass and is assumed to obey Hooke's law, i.e. the extension of the spring is proportional to the applied force within the elastic limit.

Let a mass m be attached to the lower end of the spring, causing it to extend an amount δ. In the equilibrium position, fig. 6.8(b),

downward force exerted by the mass = upward restoring force in the spring

i.e. $$mg = k\delta$$

where k = spring stiffness, with units newtons per metre (N/m).

If the mass is now moved downwards through a vertical distance x from the equilibrium position, the net force acting on the mass will be as shown in fig. 6.8(c),

i.e. $$\text{net force} = mg - (kx + k\delta)$$

$$= k\delta - kx - k\delta$$

$$= -kx$$

i.e. the net force acting on the mass is directly proportional to the displacement of the mass from the equilibrium position. The minus sign

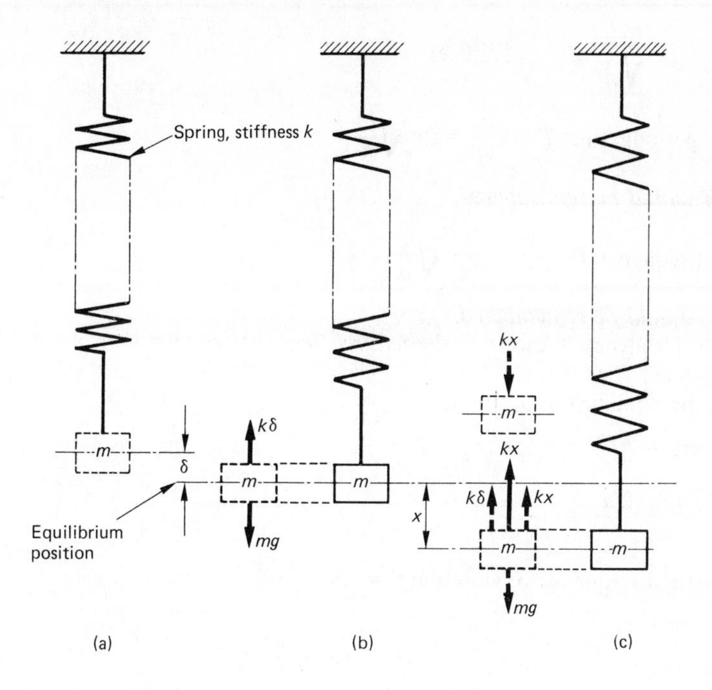

Fig. 6.8 Oscillating spring

indicates that the net force is acting towards the equilibrium position and in the opposite direction to the displacement.

If the mass is now released, it will oscillate in a vertical plane about the equilibrium position. From Newton's second law of motion, the net force will accelerate the mass in the direction of the force, i.e. towards the equilibrium position. Thus,

net force = mass x acceleration

i.e. $$-kx = ma$$

or $$a = -\frac{k}{m}x$$

which compares with $a = -\omega^2 x$ (equation (iii) in section 6.3);

i.e. the mass is oscillating about the equilibrium position with simple harmonic motion. Thus,

$$a = -\frac{k}{m}x = -\omega^2 x$$

or $$\omega^2 = \frac{k}{m}$$

$$\therefore \quad \omega = \sqrt{\frac{k}{m}} = \sqrt{\frac{\text{stiffness}}{\text{mass}}}$$

and periodic time $T = \frac{2\pi}{\omega} = 2\pi\sqrt{\frac{m}{k}}$

which should be remembered.

$$\text{Frequency } f = \frac{1}{T} = \frac{1}{2\pi}\sqrt{\frac{k}{m}}$$

which should be remembered.

This frequency is known as the *natural frequency of vibration* of the spring–mass system.

In the equilibrium position,

$$mg = k\delta$$

$$\therefore \quad k = \frac{mg}{\delta}$$

$$\therefore \quad \text{natural frequency of vibration } f = \frac{1}{2\pi}\sqrt{\frac{mg}{m\delta}}$$

or

$$f = \frac{1}{2\pi}\sqrt{\frac{g}{\delta}}$$

which it is useful to remember.

If the mass of the spring (m_s) is not negligible, then one third of this mass is added to the oscillating mass m;

$$\therefore \quad f = \frac{1}{2\pi}\sqrt{\frac{k}{m + m_s/3}}$$

which it is useful to remember. (The proof of this is beyond the scope of this book.)

Example 1 The static deflection of a spring supporting a mass is 15 mm. Determine the natural frequency of vibration of the spring–mass system.

$$f = \frac{1}{2\pi}\sqrt{\frac{g}{\delta}}$$

where $g = 9.81\text{ m/s}^2$ and $\delta = 15\text{ mm} = 0.015\text{ m}$

$$\therefore \quad f = \frac{1}{2\pi}\sqrt{\frac{9.81\text{ m/s}^2}{0.015\text{ m}}}$$

$$= 4.07\text{ Hz}$$

i.e. the natural frequency of vibration is 4.07 Hz.

Example 2 A spring of stiffness 2 kN/m and supporting a mass of 8 kg oscillates freely with an amplitude of vibration of 18 mm. Determine the frequency of the vibration and the velocity of the mass when it is 6 mm from the equilibrium position.

$$f = \frac{1}{2\pi}\sqrt{\frac{k}{m}}$$

where $k = 2\,\text{kN/m} = 2 \times 10^3\,\text{N/m}$ and $m = 8\,\text{kg}$

$$\therefore \quad f = \frac{1}{2\pi}\sqrt{\frac{2 \times 10^3\,\text{N/m}}{8\,\text{kg}}}$$

$$= 2.52\,\text{Hz}$$

i.e. the frequency is 2.52 Hz.

From section 6.3,

$$v = \omega r \sin \omega t \quad \text{(ignoring the sign)}$$

where $\omega = \sqrt{(k/m)} = 15.81$ rad/s and $r = 18\,\text{mm} = 0.018\,\text{m}$

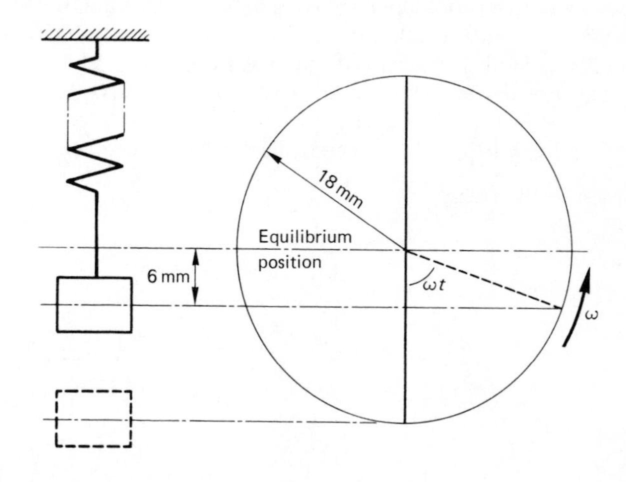

Fig. 6.9

From fig. 6.9,

$$\cos \omega t = \frac{6\,\text{mm}}{18\,\text{mm}} = 0.333$$

$$\therefore \quad \omega t = 70.53°$$

$$\therefore \quad v = 15.81\,\text{rad/s} \times 0.018\,\text{m} \times \sin 70.53°$$

$$= 0.268\,\text{m/s}$$

i.e. the velocity when the mass is 6 mm from the equilibrium position is 0.268 m/s.

Example 3 A spring of mass 6 kg oscillates with a frequency of 5 Hz when supporting a mass of 20 kg. What is the stiffness of the spring?

$$f = \frac{1}{2\pi}\sqrt{\frac{k}{m + m_s/3}}$$

$$\therefore \quad k = (2\pi f)^2 (m + m_s/3)$$

where $f = 5\,\text{Hz}$ $\quad m = 20\,\text{kg}$ $\quad$ and $\quad m_s = 6\,\text{kg}$

$$\therefore \quad k = (2\pi\,\text{rad} \times 5\,\text{Hz})^2\,[20\,\text{kg} + (6\,\text{kg})/3]$$

$$= 21.7 \times 10^3\,\text{N/m} \quad \text{or} \quad 21.7\,\text{kN/m}$$

i.e. the stiffness is 21.7 kN/m.

6.5 Oscillations of a simple pendulum

A simple pendulum is a concentrated mass suspended from a fixed point by a string which is assumed to have no mass and which will not stretch. Figure 6.10(a) shows a simple pendulum oscillating between OA and OB. At the instant shown, the angular displacement from the mean position OC is θ radians and the linear displacement of the mass from C is x. The downward force mg being exerted by the mass can be resolved into two perpendicular components as shown in fig. 6.10(b).

Referring to fig. 6.10(b), the perpendicular components of mg are

i) the tension in the string
i.e. $\quad mg \cos \theta$

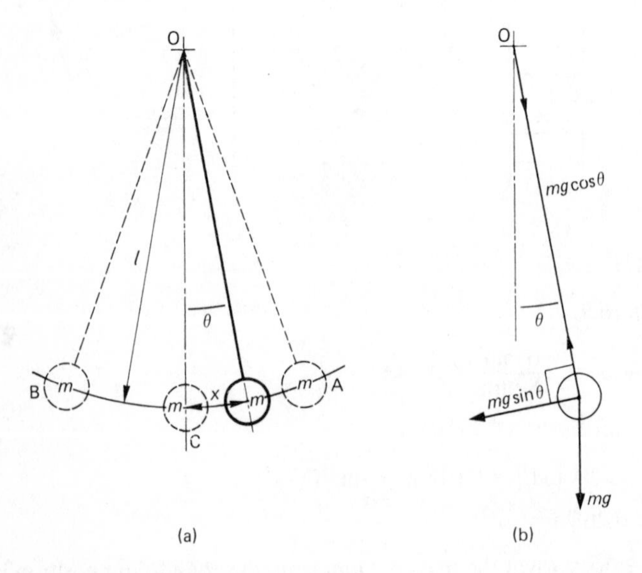

Fig. 6.10 Simple pendulum

ii) the tangential force in the direction of the mean position OC
i.e. $-mg \sin \theta$
The minus sign indicates that the force is acting in the opposite direction to the displacement for *all* positions of the pendulum between OA and OB.

The tension in the string has no effect on the motion and is thus ignored. The tangential force in the direction of OC will produce an acceleration which is *always* directed towards the mean position.

From Newton's second law of motion,

i.e. force = mass x acceleration

$$-mg \sin \theta = ma$$

or $$a = -g \sin \theta$$

For small values of θ, typically less than 10°,

$$\sin \theta \approx x/l$$

$$\therefore \quad a = -\frac{g}{l} x$$

which compares with $a = -\omega^2 x$ (equation (iii) in section 6.3);

i.e. the pendulum is oscillating between OA and OB with simple harmonic motion. Thus,

$$a = -\frac{g}{l} x = -\omega^2 x$$

or $$\omega^2 = \frac{g}{l}$$

$$\therefore \quad \omega = \sqrt{\frac{g}{l}}$$

and $$\text{periodic time } T = \frac{2\pi}{\omega} = 2\pi \sqrt{\frac{l}{g}}$$

which should be remembered.

$$\text{Frequency } f = \frac{1}{T} = \frac{1}{2\pi} \sqrt{\frac{g}{l}}$$

which should be remembered.

This frequency is also known as the natural frequency of oscillation of a simple pendulum. Notice that the periodic time and the natural frequency of oscillation of a simple pendulum are independent of the mass. It should be noted that the above equations are valid only for small amplitudes of motion, typically less than 10°.

Example 1 Determine the frequency and periodic time of a simple pendulum of length 1.3 m.

$$\text{Frequency } f = \frac{1}{2\pi}\sqrt{\frac{g}{l}}$$

where $g = 9.81\ \text{m/s}^2$ and $l = 1.3\ \text{m}$

$$\therefore\ f = \frac{1}{2\pi}\sqrt{\frac{9.81\ \text{m/s}^2}{1.3\ \text{m}}}$$

$$= 0.437\ \text{Hz}$$

i.e. the frequency is 0.437 Hz.

$$\text{Periodic time } T = 1/f$$

$$= \frac{1}{0.437\ \text{Hz}}$$

$$= 2.29\ \text{s}$$

i.e. the periodic time is 2.29 seconds.

Example 2 A simple pendulum is oscillating with an amplitude of 6°. If the pendulum is 0.9 m long, determine (a) the periodic time, (b) the maximum angular velocity, (c) the maximum angular acceleration.

a) $T = 2\pi\sqrt{(l/g)}$

where $l = 0.9\ \text{m}$ and $g = 9.81\ \text{m/s}^2$

$$\therefore\ T = 2\pi\ \text{rad}\sqrt{\frac{0.9\ \text{m}}{9.81\ \text{m/s}^2}}$$

$$= 1.9\ \text{s}$$

i.e. the periodic time is 1.9 seconds.

b) From section 6.3 and ignoring the sign,

$$v_{\text{max.}} = \omega r$$

Let Ω = maximum angular velocity of the pendulum about O; then, for a simple pendulum of length l,

$$\Omega = \frac{v_{\text{max.}}}{l} = \frac{\omega r}{l}$$

But $r/l = \theta$ (the maximum angular displacement)

$$\therefore\ \Omega = \omega\theta$$

which it is useful to remember.

Here $\omega = \dfrac{2\pi}{T} = \dfrac{2\pi\ \text{rad}}{1.9\ \text{s}} = 3.31\ \text{rad/s}$ and $\theta = 6° = 0.105\ \text{rad}$

$\therefore \quad \Omega = 3.31 \text{ rad/s} \times 0.105 \text{ rad}$

$= 0.348 \text{ rad/s}$

i.e. the maximum angular velocity is 0.348 rad/s.

c) From section 6.3 and ignoring the sign,

$$a_{\text{max.}} = \omega^2 r$$

Let α = the maximum angular acceleration of the pendulum about O; then, for a simple pendulum of length l,

$$\alpha = a_{\text{max.}}/l = \omega^2 r/l$$

As before, $r/l = \theta$

$$\therefore \quad \alpha = \omega^2 \theta$$

which it is useful to remember.

$\therefore \quad \alpha = (3.31 \text{ rad/s})^2 \times 0.105 \text{ rad}$

$= 1.15 \text{ rad/s}^2$

i.e. the maximum angular acceleration is 1.15 rad/s^2.

6.6 Resonance

All solid matter has a natural frequency of vibration. This frequency is determined by:

i) the density of the material – the greater the density, the higher the natural frequency of vibration;
ii) the shape of the material – long slender objects have a lower natural frequency of vibration than short stocky objects.

Objects will vibrate at their natural frequency of vibration if, as with the spring-mass system in section 6.4, an external force is applied which disturbs the equilibrium of the system. This external force is known as a *disturbing force*.

In the spring-mass system of section 6.4, the disturbing force was provided by moving the mass downwards, thus extending the spring. On release, the mass oscillated in a vertical plane, each succeeding amplitude of motion getting smaller until the vibration died away. If the disturbing force could be applied with a frequency equal to the natural frequency of vibration of the system – i.e. every time the mass moved down, an external force acted on it – the amplitude of the motion would get larger and larger until the spring was stretched beyond its elastic limit and eventually to failure. When the mass is vibrating in such a way, the system is said to be in *resonance* and the mass is *resonating*; i.e. *resonance will occur when the frequency of the external disturbing force is equal to the natural frequency of vibration of the system*, which should be remembered.

Resonance conditions usually occur in machines or other devices which have rotating parts. For example, when machining an awkwardly shaped casting on a lathe, it is not usually possible to achieve perfect rotational balance; thus, with every revolution of the spindle, an out-of-balance disturbing force is imparted through the lathe structure to its foundations. As the speed of rotation approaches the natural frequency of vibration of the foundation, the lathe will resonate. The effect of the resonance will clearly be seen, both in the machine-tool structure and in the poor surface finish obtained on the workpiece. Other examples include high-speed domestic spin-dryers – as the spin tub increases speed, it will invariably pass through a speed equal to the natural frequency of vibration of the tub's elastic mounting, causing it to vibrate violently for several seconds. An unbalanced car wheel will cause the vehicle to resonate when the wheel rotates at a speed equal to the natural frequency of vibration of the suspension system – this is recognised by the sudden increase in noise heard within the body of the car.

Exercises on chapter 6

1 A body moves with simple harmonic motion between two points 600 mm apart. Its velocity when passing through the mid-point is 6 m/s. Determine the periodic time and the frequency of the oscillation. [0.314 s; 3.18 Hz]

2 The amplitude of a body oscillating with simple harmonic motion is 200 mm. If the periodic time for the oscillation is 1.4 seconds, determine the maximum velocity and the maximum acceleration of the body.
[0.9 m/s; 4.03 m/s^2]

3 The maximum acceleration and periodic time of a body moving with simple harmonic motion are 5.86 m/s^2 and 0.8 seconds respectively. Determine the amplitude of the motion and the maximum velocity.
[95.1 mm; 0.75 m/s]

4 If the frequency of a body oscillating with simple harmonic motion is 3 Hz and the amplitude of the motion is 60 mm, determine the velocity and acceleration of the body when it is at a point midway between the maximum displacement and the mean position. [0.036 m/s; 0.0146 m/s^2]

5 When a body oscillating with simple harmonic motion is 20 mm from the mean position, its velocity and acceleration are 4 m/s and 1.8 m/s^2 respectively. If the periodic time of the motion is 0.9 seconds, determine the amplitude and the maximum acceleration. [0.311 m; 15.15 m/s^2]

6 Define simple harmonic motion.

A body moving with simple harmonic motion has a maximum velocity and acceleration of 5 m/s and 3 m/s^2 respectively. Determine the amplitude, frequency, and periodic time of the oscillation. [8.33 m; 0.095 Hz; 10.47 s]

7 The static deflection at the foundation of a machine tool was 2.7 mm. Determine the natural frequency of vibration of the foundation. [9.6 Hz]

8 A spring deflects 3 mm when supporting a mass of 6 kg. Determine the natural frequency of vibration and the periodic time when the spring is supporting a mass of 8 kg. [7.88 Hz; 0.127 s]

9 Determine the natural frequency of vibration when a spring of stiffness 15 kN/m supports a mass of 40 kg. [3.08 Hz]

10 A spring has a mass of 10 kg and a stiffness of 12 kN/m. If, when supporting a mass, the natural frequency of vibration is 5 Hz, determine the magnitude of the mass. [8.825 kg]

11 A spring-mass system has a periodic time of 0.3 seconds when oscillating at its natural frequency. If the mass is 15 kg, determine the spring stiffness. [6.58 kN/m]

12 Show that a spring supporting a mass will oscillate in a vertical plane with simple harmonic motion.

A spring of stiffness 35 kN/m has a natural frequency of vibration of 6 Hz when supporting a mass of 20 kg. What is the mass of the spring? [13.9 kg]

13 A spring of stiffness 20 kN/m and supporting a mass of 5 kg has an amplitude of motion of 20 mm. For the mass, determine (a) its maximum velocity and acceleration, (b) its velocity and acceleration at a point 4 mm from the outermost position. [1.26 m/s, 80 m/s^2; 0.904 m/s, 55.74 m/s^2]

14 A simple pendulum is 1.1 m long. What are its periodic time and natural frequency of oscillation? [2.1 s; 0.476 Hz]

15 Show that a simple pendulum oscillates freely with simple harmonic motion.

A simple pendulum of length 0.8 m has an amplitude of motion of 5°. Determine the maximum angular velocity and angular acceleration of the bob. [0.306 rad/s; 1.07 rad/s^2]

16 The amplitude of motion of a simple pendulum is 8°. When it is 4° from the mean position, its velocity and acceleration are 2 m/s and 2 m/s^2 respectively. Determine the length of the pendulum and the periodic time of the oscillation. [3.27 m; 3.63 s]

17 Explain the meaning of the term 'resonance' as applied to a system which may oscillate, and give *two* examples where resonance may occur.

18 Describe how resonance in a domestic spin-drying machine could be overcome.

19 A body of mass 10 kg oscillates in a straight line with simple harmonic motion. When the body is displaced 0.4 m from the central position of its motion, its velocity and acceleration are 5 m/s and 30 m/s^2 respectively. Determine (a) the frequency of the oscillation, (b) the amplitude of the oscillation, (c) the maximum velocity, (d) the maximum force acting on the body. [1.38 Hz; 0.702 m; 6.08 m/s; 526.5 N]

7 Fluids in motion

7.1 Introduction

Any liquid fluid which is in motion contains energy by virtue of its pressure and density (pressure energy), velocity (kinetic energy), and its position relative to some datum (potential energy).

7.2 Pressure energy

Let the mass of fluid between X and Y in fig. 7.1 have a total volume V_1. For flow to occur, this volume must be displaced by an equal volume entering from outside the system. If the pressure in the fluid is p_1, then the work done on the fluid *inside* the system by the incoming fluid is equal to force x distance s the fluid is displaced.

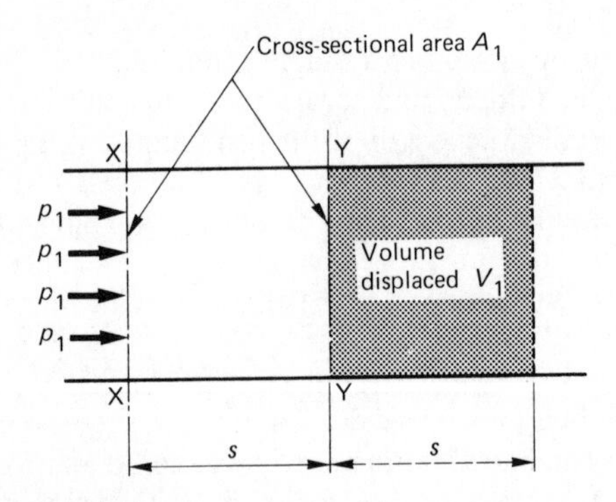

Fig. 7.1 Pressure energy

But force = pressure x cross-sectional area = $p_1 A_1$

$\therefore$ work done = $p_1 A_1 s$

But $A_1 s$ = volume displaced, V_1

$\therefore$ work done on the system = $p_1 V_1$

i.e. the incoming fluid increases the energy in the system by an amount $p_1 V_1$. This energy is known as *pressure energy*.

Similarly, the fluid leaving the system will do work equal to $p_2 V_2$ on the surroundings; i.e. the outgoing fluid reduces the energy in the system by an amount $p_2 V_2$, where p_2 and V_2 represent respectively the pressure and volume at exit from the system.

If the mass of fluid occupying volume V_1 is m_1, the mass occupying volume V_2 is m_2, and the density of the fluid is ρ (*rho*), then

$$\rho = \frac{m_1}{V_1} = \frac{m_2}{V_2}$$

Thus, at point 1 in the system, the *specific* pressure energy $= \dfrac{p_1}{\rho}$

and at point 2 in the system, the *specific* pressure energy $= \dfrac{p_2}{\rho}$

which should be remembered.

The term *specific energy* means energy per unit mass of fluid. The units for specific pressure energy are joules per kilogram (J/kg).

7.3 Kinetic energy

If a mass m of fluid has a linear velocity v, then

$$\text{kinetic energy in the fluid} = \tfrac{1}{2}mv^2$$

For unit mass of fluid,

$$\textit{specific} \text{ kinetic energy} = \tfrac{1}{2}v^2$$

which it is useful to remember.

7.4 Potential energy

If a mass m of fluid is stored at a height h above the datum, then

$$\text{potential energy stored in the fluid} = mgh$$

For unit mass of fluid,

$$\textit{specific} \text{ potential energy} = gh$$

which it is useful to remember.

7.5 Bernoulli's equation

Let unit mass of fluid flow at a steady rate through the system shown in fig. 7.2. Applying the principle of conservation of energy,

i.e. specific energy in the fluid entering the system = specific energy in the fluid leaving the system

or
$$\underbrace{\text{specific pressure energy} + \text{specific kinetic energy} + \text{specific potential energy}}_{\text{at entry}} = \underbrace{\text{specific pressure energy} + \text{specific kinetic energy} + \text{specific potential energy}}_{\text{at exit}}$$

$$\therefore \quad \frac{p_1}{\rho} + \frac{v_1^2}{2} + gh_1 = \frac{p_2}{\rho} + \frac{v_2^2}{2} + gh_2 = \text{constant}$$

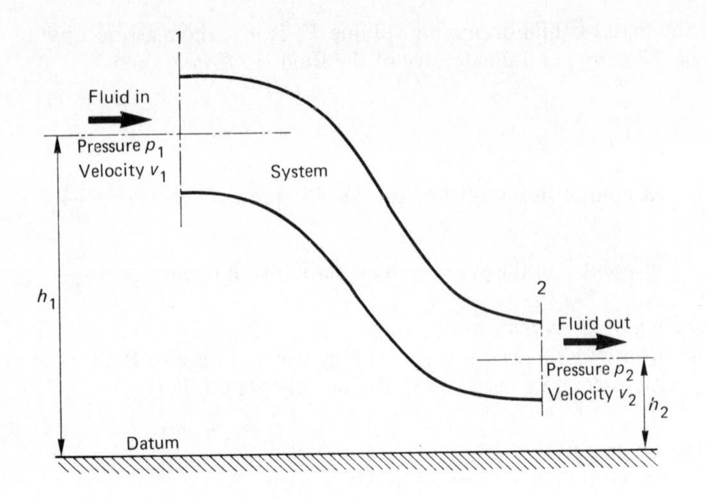

Fig. 7.2 Bernoulli's equation

Dividing both sides of the equation by g,

$$\frac{p_1}{\rho g} + \frac{v_1^2}{2g} + h_1 = \frac{p_2}{\rho g} + \frac{v_2^2}{2g} + h_2 = \text{constant}$$

This is known as Bernoulli's equation and should be remembered.

Each quantity in the Bernoulli equation is measured in terms of 'head of liquid', i.e. height of liquid above a given datum. The unit for each quantity is the *metre* (m).

Example Water enters a pipeline with a pressure of 2 bar and velocity 5 m/s. If the pressure in the water 25 m below the entry point is 1.2 bar, calculate the velocity. Take the density of water as 1000 kg/m³.

Using Bernoulli's equation,

i.e. $$\frac{p_1}{\rho g} + \frac{v_1^2}{2g} + h_1 = \frac{p_2}{\rho g} + \frac{v_2^2}{2g} + h_2$$

gives $$v_2 = v_1 + \sqrt{\left(\frac{2(p_1 - p_2)}{\rho} + (h_1 - h_2)2g\right)}$$

For convenience, make the point 25 m below the entry point the datum,

i.e. $h_2 = 0$ and $h_1 = 25$ m

Also, $v_1 = 5$ m/s $\quad p_1 = 2$ bar $= 2 \times 10^5$ N/m²

$p_2 = 1.2$ bar $= 1.2 \times 10^5$ N/m² $\quad g = 9.81$ m/s²

and $\rho = 1000$ kg/m³

$$\therefore \quad v_2 = 5\text{ m/s} + \sqrt{\left(\frac{2(2-1.2)\times 10^5\text{ N/m}^2}{1000\text{ kg/m}^3} + (25\text{ m} - 0)\times 2\times 9.81\text{ m/s}^2\right)}$$

$$= 5\text{ m/s} + 25.5\text{ m/s}$$

$$= 30.5\text{ m/s}$$

i.e. the velocity of the water is 30.5 m/s.

7.6 Frictional resistance to flow

Bernoulli's equation assumes there to be no frictional resistance to the flow. In practice, frictional resistance to flow is *always* present and reduces the available energy in the fluid at exit from the system;

i.e. specific energy in fluid entering the system = specific energy in fluid leaving the system + specific energy to overcome frictional resistance to flow

Let h_f = frictional resistance 'head'

then $$\frac{p_1}{\rho g} + \frac{v_1^2}{2g} + h_1 = \frac{p_2}{\rho g} + \frac{v_2^2}{2g} + h_2 + h_f$$

which it is useful to remember.

Example Water of density 1000 kg/m^3 enters a horizontal pipeline with a pressure and velocity of 2 MPa and 7.5 m/s respectively, and leaves with a pressure of 1.65 MPa and velocity 25 m/s. Determine the loss of head due to frictional resistance.

Frictional resistance head = energy input head – energy output head

or $$h_f = \frac{(p_1 - p_2)}{\rho g} + \frac{(v_1^2 - v_2^2)}{2g} + (h_1 - h_2)$$

where $p_1 = 2\text{ MPa} = 2\times 10^6\text{ N/m}^2$ $\quad p_2 = 1.65\text{ MPa} = 1.65\times 10^6\text{ N/m}^2$

$\rho = 1000\text{ kg/m}^3$ $\quad g = 9.81\text{ m/s}^2$ $\quad v_1 = 7.5\text{ m/s}$

and $v_2 = 25\text{ m/s}$

Since the pipeline is horizontal, $h_1 = h_2 = 0$

$$\therefore \quad h_f = \frac{(2-1.65)\times 10^6\text{ N/m}^2}{1000\text{ kg/m}^3\times 9.81\text{ m/s}^2} + \frac{(7.5^2 - 25^2)(\text{m/s})^2}{2\times 9.81\text{ m/s}^2} + 0$$

$$= 35.68\text{ m} + (-29\text{ m})$$

$$= 6.68\text{ m}$$

i.e. the loss of head due to friction is 6.68 m.

7.7 Continuity equation

Consider sections 1 and 2 in the tapered pipe shown in fig. 7.3, which is full of steadily flowing fluid. At section 1, the cross-sectional area is A_1 and the velocity of the fluid is v_1; at section 2, the area and velocity are A_2 and v_2 respectively.

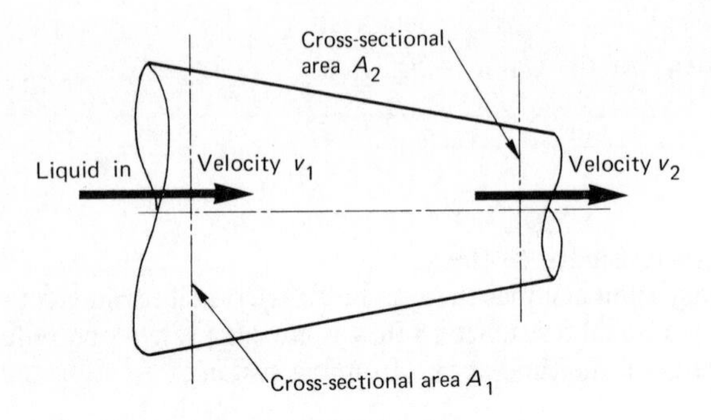

Fig. 7.3 Continuity equation

Since the pipe is full of fluid and the flow is steady,

$$\text{volume of fluid passing section 1 per unit time} = \text{volume of fluid passing section 2 per unit time}$$

or
$$A_1 v_1 = A_2 v_2$$

(since volume per unit time = area x velocity)
This is known as the continuity equation for steady flow, and should be remembered.

If the pipe is circular, with diameters d_1 and d_2 at sections 1 and 2 respectively, then

$$\frac{\pi}{4} d_1^2 v_1 = \frac{\pi}{4} d_2^2 v_2$$

or
$$d_1^2 v_1 = d_2^2 v_2$$

which it is useful to remember.

Example At a section in a tapered pipe, the diameter is 0.7 m and the velocity of the liquid is 5 m/s. Assuming the flow of the fluid to be steady, determine the velocity when the pipe diameter is 0.5 m.

Since the pipe is circular,

$$d_1^2 v_1 = d_2^2 v_2$$

$$\therefore \quad v_2 = (d_1/d_2)^2/v_1$$

where $d_1 = 0.7\text{ m}$ $\quad d_2 = 0.5\text{ m}$ and $v_1 = 5\text{ m/s}$

$$\therefore \quad v_2 = \left(\frac{0.7\text{ m}}{0.5\text{ m}}\right)^2 \times 5\text{ m/s}$$

$$= 9.8\text{ m/s}$$

i.e. the velocity is 9.8 m/s.

7.8 Applications of the Bernoulli and continuity equations

Example 1 Oil of density 800 kg/m³ enters a 500 mm diameter pipeline with a velocity of 3 m/s and pressure 400 kPa, and is discharged through an orifice 150 mm in diameter and 30 m below the entry point. If the frictional losses are equivalent to a head of 3.5 m of the oil, determine the velocity and pressure of the oil at the point of discharge.

To find the velocity, use the continuity equation for a circular pipe,

i.e. $d_1^2 v_1 = d_2^2 v_2$

$\therefore \quad v_2 = (d_1/d_2)^2 v_1$

where $d_1 = 500$ mm $\quad d_2 = 150$ mm $\quad$ and $\quad v_1 = 3$ m/s

$$\therefore \quad v_2 = \left(\frac{500\text{ mm}}{150\text{ mm}}\right)^2 \times 3\text{ m/s}$$

$$= 33.3\text{ m/s}$$

i.e. the velocity at discharge is 33.3 m/s.

To find the pressure, use Bernoulli's equation,

i.e. $$\frac{p_1}{\rho g} + \frac{v_1^2}{2g} + h_1 = \frac{p_2}{\rho g} + \frac{v_2^2}{2g} + h_2 + h_f$$

$$\therefore \quad p_2 = p_1 + \tfrac{1}{2}(v_1^2 - v_2^2)\rho + (h_1 - h_2 - h_f)\rho g$$

where $p_1 = 400$ kPa $= 400 \times 10^3$ N/m² $\quad v_1 = 3$ m/s

$v_2 = 33.3$ m/s $\quad \rho = 800$ kg/m³ $\quad g = 9.81$ m/s²

$h_f = 3.5$ m $\quad$ and, making the orifice the datum, $\quad h_2 = 0$

and $\quad h_1 = 30$ m

$$p_2 = [400 \times 10^3\text{ N/m}^2] + [\tfrac{1}{2}(3^2 - 33.3^2)(\text{m/s})^2 \times 800\text{ kg/m}^3]$$
$$+ [(30 - 0 - 3.5)\text{m} \times 9.81\text{ m/s}^2 \times 800\text{ kg/m}^3]$$
$$= [400 \times 10^3\text{ N/m}^2] + [-440 \times 10^3\text{ N/m}^2] + [208 \times 10^3\text{ N/m}^2]$$
$$= 168 \times 10^3\text{ N/m}^2 \quad \text{or} \quad 168\text{ kPa}$$

i.e. the pressure at discharge is 168 kPa.

Example 2 Water contained in an open tank is discharged to the atmosphere through an orifice 25 mm in diameter. If the surface of the water is maintained at a constant 1.6 m above the centre of the orifice, and ignoring losses, determine (a) the velocity of the water as it discharges through the orifice, (b) the volume of water flowing in litres per minute.

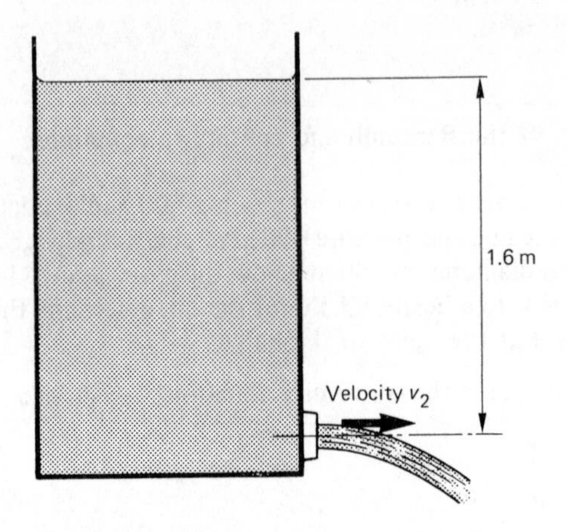

Fig. 7.4

a) Referring to fig. 7.4 and applying Bernoulli's equation,

$$\frac{p_1}{\rho g} + \frac{v_1^2}{2g} + h_1 = \frac{p_2}{\rho g} + \frac{v_2^2}{2g} + h_2 + h_f$$

Putting the datum through the centre of the orifice makes $h_2 = 0$. Also,

$p_1 = p_2$

$v_1 = 0$ (since there is no movement at the surface of the water)

and $h_f = 0$

$\therefore \; h_1 = v_2^2/2g$

or $v_2 = \sqrt{2gh_1}$

where $h_1 = 1.6\,\text{m}$ and $g = 9.81\,\text{m/s}^2$

$\therefore \; v_2 = \sqrt{(2 \times 9.81\,\text{m/s}^2 \times 1.6\,\text{m})}$

$= 5.6\,\text{m/s}$

i.e. the velocity is 5.6 m/s.

b) Volume of water flowing per second = area of orifice x velocity

$= (\pi/4) \times (0.025\,\text{m})^2 \times 5.6\,\text{m/s}$

$= 0.0027\,\text{m}^3/\text{s}$

But $1\ m^3 = 1000$ litres

$\therefore$ flow rate in litres per minute $= 0.0027\ m^3/s \times 1000\ l/m^3 \times 60\ s/min$

$= 162\ l/min$

i.e. the volume of water flowing is 162 litres per minute.

Example 3 Water with a gauge pressure of 500 kPa and having negligible velocity is discharged to the atmosphere through a horizontal nozzle 20 mm in diameter. Ignoring losses, determine the velocity of discharge and the volumetric flow rate in litres per minute. The density of the water is 1000 kg/m³.

From Bernoulli's equation,

$$\frac{p_1}{\rho g} + \frac{v_1^2}{2g} + h_1 = \frac{p_2}{\rho g} + \frac{v_2^2}{2g} + h_2 + h_f$$

Here $p_2 = 0 \quad v_1 = 0 \quad h_f = 0$

and $h_1 = h_2 = 0$ (since nozzle is horizontal)

$\therefore\ p_1/\rho g = v_2^2/2g$

or $v_2 = \sqrt{(2p_1/\rho)}$

where $p_1 = 500\ kPa = 500 \times 10^3\ N/m^2$ and $\rho = 1000\ kg/m^3$

$$\therefore\ v_2 = \sqrt{\frac{2 \times 500 \times 10^3\ N/m^2}{1000\ kg/m^3}}$$

$$= 31.62\ m/s$$

i.e. the velocity of discharge is 31.62 m/s.

Volumetric flow rate $= Av$

where $A = (\pi/4) \times (0.02\ m)^2 = 0.000\ 314\ m^2$

$\therefore$ volumetric flow rate $= 0.000\ 314\ m^2 \times 31.62\ m/s$

$= 0.009\ 93\ m^3/s$

$= 596$ litres/min

(since $1\ m^3/s = 60 \times 10^3$ litres/min)

i.e. the volumetric flow rate is 596 litres/min.

Example 4 The outlet pipe from a reservoir serving a hydroelectric power station is 30 m below the surface of the water. The pipe is 400 m long and is inclined at 60° to the horizontal. The outlet nozzle at the turbine is 600 mm in diameter. Taking frictional losses as equivalent to 2% of the potential head, determine (a) the pressure in the water at entry to the pipe, (b) the velocity of the water issuing from the nozzle, (c) the velocity in the pipe when its diameter is 2 m. Take the density of water as 1000 kg/m^3.

a) Pressure in the water at entry to the pipe $= \rho g h$

where $\rho = 1000\,\text{kg/m}^3$ $\quad g = 9.81\,\text{m/s}^2$ $\quad$ and $\quad h = 30\,\text{m}$

$\therefore$ pressure $= 1000\,\text{kg/m}^3 \times 9.81\,\text{m/s}^2 \times 30\,\text{m}$

$= 294.3 \times 10^3\,\text{N/m}^2$ or $294.3\,\text{kPa}$

i.e. the pressure at entry to the pipe is 294.3 kPa.

b) From Bernoulli's equation,

$$\frac{p_1}{\rho g} + \frac{v_1^2}{2g} + h_1 = \frac{p_2}{\rho g} + \frac{v_2^2}{2g} + h_2 + h_f$$

Referring to fig. 7.5 and making the nozzle exit the datum,

$h_2 = 0 \qquad h_1 = (400\,\text{m} \times \sin 60°) + 30\,\text{m} = 376.4\,\text{m}$

and $h_f = 2\%$ of $h_1 = 7.53\,\text{m}$

Also, $p_1 = p_2$ (since the surface of the water and the exit from the nozzle are both at atmospheric pressure) and $v_1 = 0$

$\therefore$ exit velocity $v_2 = \sqrt{[2g(h_1 - h_f)]}$

$= \sqrt{[(2 \times 9.81\,\text{m/s}^2)(376.4\,\text{m} - 7.53\,\text{m})]}$

$= 85.07\,\text{m/s}$

i.e. the velocity of the water issuing from the nozzle is 85.07 m/s.

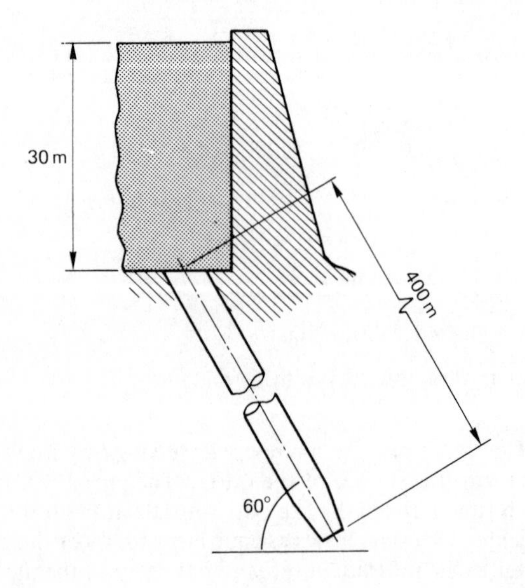

Fig. 7.5

c) Using the continuity equation for a circular pipe,

i.e. $d_1^2 v_1 = d_2^2 v_2$

$v_1 = (d_2/d_1)^2 v_2$

where $d_1 = 2\,\text{m}$ and $d_2 = 600\,\text{mm} = 0.6\,\text{m}$

$$\therefore \quad v_1 = \left(\frac{0.6\,\text{m}}{2\,\text{m}}\right)^2 \times 85.07\,\text{m/s}$$

$$= 7.66\,\text{m/s}$$

i.e. the velocity at the point where the pipe is 2 m diameter is 7.66 m/s.

7.9 Momentum of a jet

Momentum is the product of mass m and velocity v,

i.e. momentum = mass x velocity

or momentum = mv

which should be remembered.

The unit for momentum is the kilogram metre per second (kg m/s).

Let a jet of liquid of density ρ (*rho*) and flowing with velocity v have a cross-sectional area A. The volumetric flow rate, i.e. the volume of liquid flowing per unit time, is then given by

volumetric flow rate = area of jet x velocity

$= Av$

and the mass flow rate, i.e. the mass of liquid flowing per unit time, is given by

mass flow rate = density x volumetric flow rate

$= \rho Av$

$\therefore$ momentum of the liquid per unit time = mass flow rate x velocity

$= \rho Av.v$

$= \rho Av^2$

which should be remembered.

Example A jet of water has a diameter of 0.04 m and velocity 30 m/s. If the density of the water is 1000 kg/m³, calculate the momentum per second in the jet.

Momentum per unit time = ρAv^2

where $\rho = 1000\,\text{kg/m}^3$ $\quad A = (\pi/4) \times (0.04\,\text{m})^2 = 0.0013\,\text{m}^2$

and $v = 30\,\text{m/s}$

$\therefore$ momentum per unit time $= 1000\,\text{kg/m}^3 \times 0.0013\,\text{m}^2 \times (30\,\text{m/s})^2$

$= 1170\,\text{kg m/s per second}$

i.e. the momentum per second in the jet is 1170 kg m/s.

7.10 Impulse

An impulse is the product of force F and time t and is equal to the change in momentum (see section 5.4, *Engineering science for technicians vol. 2*, by I. McDonagh, G. Waterworth, and R.P. Phillips),

i.e. impulse = change in momentum

or $$Ft = m(v-u)$$

where v and u are the final and initial velocities respectively,

$$\therefore \quad F = \frac{m(v-u)}{t}$$

i.e. force = change in momentum per unit time in the same direction as the change

7.11 Force of a jet on a flat plate

If a jet of liquid impinges on a flat plate as shown in fig. 7.6, the liquid will disperse in radial directions, reducing the velocity of the jet in the original direction to zero; i.e., upon impact with the flat plate, the momentum of the liquid in the original direction is zero.

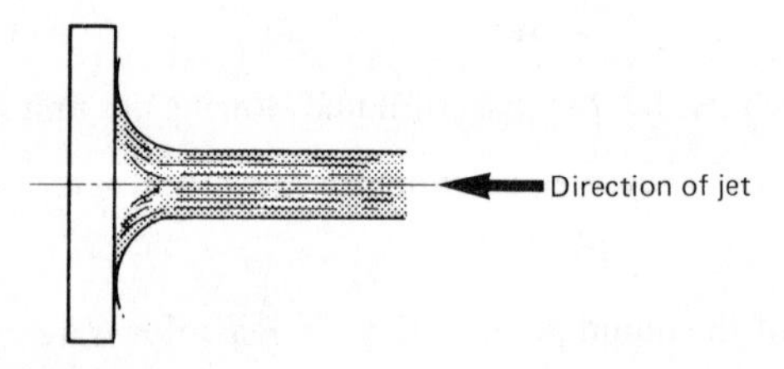

Fig. 7.6 Force of a jet on a flat plate

$$\text{Change in momentum per unit time} = \text{final momentum} - \text{initial momentum}$$

$$= 0 - \rho A v^2$$

$$= -\rho A v^2$$

The minus sign indicates that the direction of the change is in the opposite direction to the incoming jet.

But, change in momentum per unit time = force in the direction of the change

From Newton's third law of motion, i.e. to every force there is an equal and opposite force reacting, the *reaction* to the change in momentum per unit time will be the force on the plate,

i.e. force on the plate = $\rho A v^2$

and is in the *same* direction as the incoming jet, *which should be remembered.*

Example Oil of density 800 kg/m^3 impinges on a flat plate with a velocity of 6 m/s. If the diameter of the jet is 20 mm, determine the force on the plate.

Force on the plate = $\rho A v^2$

where ρ = 800 kg/m^3 $\quad A = (\pi/4) \times (0.02\text{ m})^2 = 0.000\,314\text{ m}^2$

and v = 6 m/s

$\therefore$ force on the plate = $800\text{ kg/m}^3 \times 0.000\,314\text{ m}^2 \times (6\text{ m/s})^2$

= 9 N

i.e. the force on the plate is 9 N.

Exercises on chapter 7

In the following exercises, the density of water is 1000 kg/m^3.

1 At a point in a horizontal pipeline, the water pressure is 2.2 bar and the velocity is 5 m/s. What is the specific energy in the water? [232.5 J/kg]
2 Write down, in symbolic form, Bernoulli's equation in terms of 'head' of liquid, stating the name of and the units for each symbol.
Oil is stored at a height of 6 m above the outlet nozzle. If the velocity of flow at the outlet is 9 m/s, determine the frictional resistance to flow in terms of head of oil. [1.87 m]
3 The pressure and velocity of a liquid at a point in a pipeline were found to be 3 bar and 17.5 m/s respectively. At a second point in the line, 5 m above the first point, the pressure and velocity were 2.8 bar and 14 m/s respectively. If the relative density of the liquid is 1.2, find the loss of head. [2.32 m]
4 Water with a velocity of 10 m/s enters a pipeline at a point 50 m above the outlet. If the diameter of the outlet pipe is 0.2 m and the loss of head due to friction is 5.5 m, calculate the inlet diameter of the pipe. [0.353 m]
5 In a pumped-storage hydroelectric scheme, water is raised 300 m from the lower to the upper reservoir. If the frictional losses are equivalent to a head of 35 m, find the pressure in the pump required to maintain a volumetric flow rate of 270 000 litres per minute through a 0.8 m diameter exit port at the pump. (1 litre = 10^{-3} m^3.) [2.89 MN/m^2]
6 The surface of the water in a supply chamber is maintained at a constant height of 3.5 m above the outlet orifice, which is 10 mm in diameter. If there are no frictional losses, determine the volume of water issuing from the orifice in litres per minute. [39.1 litres/min]

7 It is required to spray water on to a fire from a height of 12 m above the fire-hose nozzle. Determine the velocity of the water issuing from the nozzle. If frictional losses cause a pressure drop of 0.2 bar between the pump and the nozzle outlet, what is the pump pressure? [15.34 m/s; 1.38 bar]

8 Water leaves a pump at a pressure of 2.5 bar and issues through a nozzle 15 mm in diameter at a distance of 150 m from the pump. If the pressure drop in the pipeline is 0.02 kN/m^2 per metre, calculate (a) the velocity of the water at exit from the nozzle, (b) the force the water jet would exert on a flat surface perpendicular or normal to the jet. [22.23 m/s; 87.3 N]

9 A jet of water of diameter 40 mm impinges on a flat plate with a force of 200 N. If the plate is normal to the jet, what is the velocity of the water? [12.6 m/s]

10 A water jet flowing at the rate of 200 litres per minute strikes a flat plate, exerting on it a force of 50 N. What is the velocity of the water? [15 m/s]

11 A jet of water impinges normally on a fixed flat plate with a force of 600 N. If the velocity of the water on impact is 15 m/s, calculate the flow of water in litres per minute. [2400 litres/min]

12 Water, issuing from a nozzle 20 mm in diameter, impinges normally on a fixed flat plate. If the water is supplied from a container which is maintained at a constant height of 20 m above the nozzle, calculate the force on the plate. Neglect frictional losses. [123.3 N]

13 Water enters a pipeline with a pressure of 300 kPa and negligible velocity and leaves through a nozzle 100 mm in diameter, 20 m below the entry point. If the frictional losses are equivalent to a head of 6.5 m, determine the force the water will exert on a flat plate which is normal to the nozzle exit. [6793 N]

14 A pipeline, full of water, tapers from 0.7 m diameter at point A to 0.35 m diameter at point B which is 25 m vertically below A. If at point A the pressure and velocity are 600 kPa and 5 m/s respectively, calculate the velocity at B. If the frictional head loss is 3.6 m, what will be the pressure at B? [20 m/s; 622.4 kPa]

15 Water flows through a horizontal pipeline at the rate of 785.4 kg/s. If the mean velocity of the flow is 4 m/s, what is the diameter of the pipe? If the pipeline is 3 km long, determine the pump pressure required to maintain the flow rate if the frictional losses amount to 0.002 m per metre run of pipeline. [0.5 m; 58.86 kN/m^2]

Index

Mechanical science for technicians